T0113085

# HOW INSECTS WORK

How Insects Work:
*An Illustrated Guide to the Wonders of Form and Function—from Antennae to Wings*

Page 224 is a continuation of this copyright page.

Published in North America by The Experiment, LLC, in 2020.



The Experiment, LLC | 220 East 23rd Street, Suite 600 | New York, NY 10010-4658
theexperimentpublishing.com



Library of Congress Cataloging-in-Publication Data

Names: Taylor, Marianne, 1972- author.
Title: How insects work : an illustrated guide to the wonders of form
and function-from antennae to wings / Marianne Taylor.
Other titles: How nature works.
Description: New York City : The Experiment, 2020. | Series: How nature works | Includes index.
Identifiers: LCCN 2019056118 (print) | LCCN 2019056119 (ebook)
ISBN 9781615196494 | ISBN 9781615196500 (ebook)
Subjects: LCSH: Insects--Anatomy.
Classification: LCC QL494 .T39 2020 (print) | LCC QL494 (ebook) | DDC 595.7--dc23
LC record available at https://lccn.loc.gov/2019056118
LC ebook record available at https://lccn.loc.gov/2019056119

ISBN 978-1-61519-649-4
Ebook ISBN 978-1-61519-650-0

Conceived, designed, and produced by UniPress Books Limited
unipressbooks.com

Cover design by Beth Bugler

Manufactured in China

First printing April 2020
10 9 8 7 6 5 4 3 2 1

# HOW INSECTS WORK

## AN ILLUSTRATED GUIDE TO THE WONDERS OF **FORM AND FUNCTION**

FROM ANTENNAE TO WINGS

MARIANNE TAYLOR

NEW YORK

# CONTENTS

Introduction 6

1 ANCESTORS AND EVOLUTION 8

1.1 Rise of the arthropods
1.2 The first insects?
1.3 The first fliers
1.4 Insect evolutionary tree
1.5 Carboniferous Earth: A land of giants
1.6 Recent evolution

2 THE INSECT BODY PLAN 32

2.1 Exoskeleton
2.2 The three body sections
2.3 Segmentation and appendages
2.4 Limb structure
2.5 The wings and elytra
2.6 Unusual bodies

3 THE SENSES AND THE NERVOUS SYSTEM 48

3.1 Eyes
3.2 Antennae
3.3 Chemoreception
3.4 Hearing, touch, and more
3.5 Brains, ganglia, and nerves
3.6 Insect intelligence

4 MOVEMENT 64

4.1 Muscular system
4.2 Movement on land
4.3 Flight
4.4 Swimming and diving
4.5 Escaping danger
4.6 Immobility

5 FEEDING AND DIGESTION 78

5.1 Mouthpart anatomy
5.2 Types of diet
5.3 The digestive tract
5.4 Processing food
5.5 Changes during life cycle
5.6 Drinking and fluid balance

6 THE RESPIRATORY AND CIRCULATORY SYSTEMS 92

6.1 Breathing system
6.2 Gas exchange
6.3 Circulatory system
6.4 Hemolymph
6.5 Unusual adaptations
6.6 Hormones

7 THE REPRODUCTIVE SYSTEM 106

7.1 Male reproductive anatomy
7.2 Female reproductive anatomy
7.3 Mating and fertilization
7.4 Parthenogenesis
7.5 Laying eggs
7.6 Unusual adaptations

8 EGGS AND LARVAE 126

8.1 Types of eggs
8.2 Development in the egg
8.3 Types of larvae
8.4 Feeding
8.5 Growth and molt
8.6 Lifestyle changes

9 METAMORPHOSIS 140

9.1 Types of life cycle
9.2 Incomplete metamorphosis
9.3 Full metamorphosis
9.4 Transformation within the pupa
9.5 Emergence
9.6 Maturation in adulthood

10 BEHAVIOR AND ANATOMY 156

10.1 Feeding behavior
10.2 Breeding behavior
10.3 Parental care
10.4 Seasonal behavior
10.5 Eusocial insects
10.6 Interspecies interactions

11 CELLS AND BIOCHEMISTRY 178

11.1 Structure of a typical cell
11.2 Cell organelles
11.3 Cell replication
11.4 Immunology
11.5 Specialized cell types
11.6 Insects in cellular research

12 DIVERSITY AND CONSERVATION 192

12.1 Types of insects
12.2 Insect communities in different habitats
12.3 Record-breakers
12.4 Threats facing insects
12.5 Extinction
12.6 Insect conservation

Glossary 216
Index 218
Credits 224

# INTRODUCTION

**Across the great and varied sweep of life on Earth, insects stand out as one of the greatest success stories. Most invertebrate animals live in the oceans and fresh waters, but insects have truly conquered the land, and (as the only winged invertebrates that have ever lived) they have also mastered the air. This mastery comes courtesy of a basic anatomy that meets the challenges of life out of water, and thanks also to countless anatomical modifications that allow insects to thrive in so many different habitats and ecological niches.**

For all their fabulous variety, insects have the same fundamental body-plan, which allows them to be recognised at a glance. The segmented body has three distinct sections: the head, the thorax, and the abdomen. There are six legs and (usually) two pairs of wings attached to the thoracic segments, and there are obvious eyes and various sensory and feeding appendages on the head. Whether the insect crawls, runs, climbs, or hangs, whether it flies with a rattling zoom, a buzz, or a flutter, whether it hunts prey, chews leaves, or sucks nectar (or blood), it does so with its own version of the same physical equipment that first evolved more than 350 million years ago.

One of the keys to insects' success in the open air lies in their outer covering—a waxy cuticle that helps prevent their tiny bodies from dehydrating. To take oxygen from the air, they use spiracles—breathing apertures in the body segments, which take in air passively and can be opened and closed as needed. Instead of blood contained in vessels, they have free-flowing hemolymph, which helps keep their bodies rigid and aids movement, as well as transporting nutrients and waste materials to the appropriate parts of the body. The nervous system is

Ⓥ A spectacular Spiny Flower Mantis exhibits its eye-spot wing markings, to startle a predator.

modular—in a sense, each of the body segments has its own individual and autonomous brain—and some other body systems show a similar modularization. These are just a few of the many ways in which insect bodies are structured and function completely differently to our own, though it is the process of complete bodily metamorphosis, from wormlike larva to winged adult, that astounds us most of all.

(A) The male Stag Beetle is a fighting machine—his huge jaws are for wrestling a rival rather than biting.

(V) The Madagascan Sunset Moth is noted for its long migrations as well as its dazzling colors.

## ENDLESS FORMS MOST BEAUTIFUL

There are well over a million species of insects known to science today, and probably many more that are as yet undiscovered. They have adapted to live on every continent and in every kind of environment, fulfilling an array of ecological roles. Some are our constant companions—a few are even domesticated for our use—and others are our sworn enemies, but the vast majority are almost unknown to most of us. This book aims to unravel the mystique of insects—how their bodies work, how they lead their lives, and how deeply and completely the lives of other organisms on Earth (including ourselves) depend on them.

1

# ANCESTORS AND EVOLUTION

Insects are the most successful and diverse of all land animals. Their long evolutionary history has furnished them with uniquely adaptable bodies, capable of thriving in all environments. They have survived five mass extinctions and many species have even surmounted the greatest challenge of all—successfully surviving alongside humanity.


1.1 · Rise of the arthropods

1.2 · The first insects?

1.3 · The first fliers

1.4 · Insect evolutionary tree

1.5 · Carboniferous Earth: A land of giants

1.6 · Recent evolution


Trilobites were ancient arthropods and cousins to the insects—their segmented structure has proved an enduringly successful body plan.

# RISE OF THE ARTHROPODS

**Insects are invertebrate animals belonging to the larger group Arthropoda, meaning "jointed feet." Their bodies also have movable joints at certain points on their rigid outer shell.**

Humans, along with other mammals and all vertebrates, have an endoskeleton. The strong, rigid framework of bones is inside, with muscles, blood vessels, nerves, and other soft, fleshy structures around them. In insects, this anatomy is reversed. Their structural framework, the exoskeleton, is on the outside and the soft parts are within.

As well as insects, the arthropod group today includes crabs and other crustaceans, spiders and other arachnids, and the many-legged millipedes and centipedes. Their basic body-plan is bilateral symmetry, with left and right as mirror images, and the body is divided into many sections, called segments. The jointed legs and other limbs or appendages are in pairs along each side.

## THE EXOSKELETON

The first animal with a segmented exoskeleton and jointed limbs lived more than 540 million years ago, in the sea. Its appearance was a major event in evolution, introducing a very adaptable body arrangement. (An incidental benefit is that a strong, tough exoskeleton was much more likely to be preserved as a fossil—a huge help for our studies today.) The earliest fossil arthropods were small organisms, from the size of rice grains to grapes, and resembled a mix of worm and crustacean. But these remains are vague and difficult to classify.

Fossil trilobites: These early arthropods first appeared on Earth some 521 million years ago.

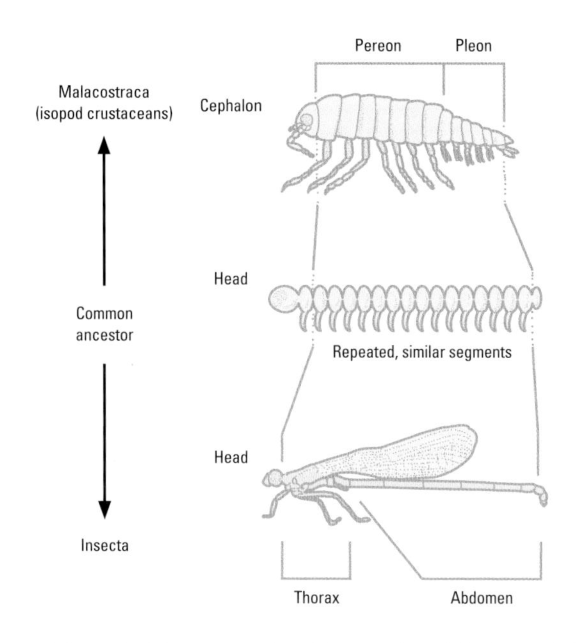


Modern arthropods like crustaceans (here an isopod) and insects (here a damselfly) both descend from an animal with simple repeated body segments, each bearing jointed appendages.

Trilobites were swimming arthropods that lived in seas worldwide some 525 million years ago, during the Cambrian Period. The sea scorpions, or eurypterids, formed another prominent arthropod group. These fearsome predators first lived around 460 million years ago, in the Ordovician Period. Some, such as *Jaekelopterus*, were among the largest arthropods of all time, at 8 feet (2.5m) long. Both trilobites and eurypterids flourished for more than 200 million years, then gradually declined. The last of them died out in the Permian–Triassic mass extinction event, 252 million years ago.

**ONTO LAND**

Meanwhile, other arthropods had moved onto land, probably 430–450 million years ago during the Ordovician Period. Their basic body design and anatomy were already well suited to terrestrial life—a feature in evolutionary science known as preadaptation. The tough, impermeable exoskeleton greatly reduced body fluid loss in air, and the jointed limbs held up the body for moving around. Some of the first terrestrial arthropods were trigonotarbids, which were arachnids similar to their modern spider relatives. Mostly small, ⅕–⅘ inch (0.5–2cm) in length, they went extinct about 290–300 million years ago. A very different destiny awaited one of the next arthropod groups to colonize land—the insects.

# THE FIRST INSECTS?

**In the Devonian Period, 359–419 million years ago, insects were spreading into early terrestrial habitats, helped along by pioneer plants, which established environments and food sources.**

Ⓐ Silverfish have altered little in anatomy since their ancestors first appeared about 380 million years ago.

A single fossil, as small as this "o," may be the oldest known insect fossil. The fossil in question is *Rhyniognatha hirsti* and is the subject of continuing debate. It consists of most of a tiny head but little else. It was excavated in 1919 from flint-like rocks called Rhynie chert, at a village near Aberdeen, Scotland. In 1926 it was interpreted as an arthropod, but as a springtail or collembolan, *Rhyniella praecursor*, a relative of true insects. In 1928, reexamination, especially of the mandibles—insect "jaws"—showed it to be a true insect and it was renamed *Rhyniognatha hirsti*. The specimen is dated to the early Devonian Period, 395–400 million years ago.

## CHANGING OPINIONS

For decades *Rhyniognatha* was famous as the very earliest insect fossil. In 2002 it was reexamined using much improved microscopes. The verdict was that its mandibles moved with a scissor-like action, using two joints. This anatomy is found in the mandibles of flying insects, but not in non-fliers, which were assumed to have evolved first. The earliest fossils of flying insects do not appear until at least 70–80 million years after *Rhyniognatha*. So if *Rhyniognatha* was indeed a flier, this would push back the origins of insect flight many millions of years, and the beginnings of insects in general even further.

The *Rhyniognatha* specimen itself showed no wings, but this is unsurprising since it was originally preserved in a hot spring where tiny fragile wings were unlikely to survive. These 2002 studies suggested the whole creature resembled a mayfly that could rest on a human fingernail.

## NOT AN INSECT?

In 2017 there was another twist. The *Rhyniognatha* specimen was examined again using more advanced microscopes, which revealed previously overlooked fragments including additional mouthparts. The new conclusion was that *Rhyniognatha* was not an insect. The anatomy of the head parts, plus

➢ Firebrats are widespread primitive insects which often occur in hot, humid environments.

information about other arthropods present in the fossil-dig location at the time, meant it was more likely an early relative of centipedes. Perhaps future generations of microscopes and experts will settle the status of *Rhyniognatha*.

Fossils from about 380 million years ago onward, in eastern North America and gradually spreading to Eurasia, are generally accepted as early insects. They were wingless and from the groups known as Archaeognatha (rock and jumping bristletails), and Zygentoma (silverfish and firebrats). These kinds of insects are known as "primitive," which in biology means a form that arose early in evolutionary history and has changed or been modified little since.

It is not the case that evolution has passed them by, merely that environmental conditions that suit their survival have remained in more or less the same state for all those millennia. They have continued to evolve, but only to become better and better at living in the same way that they always have.

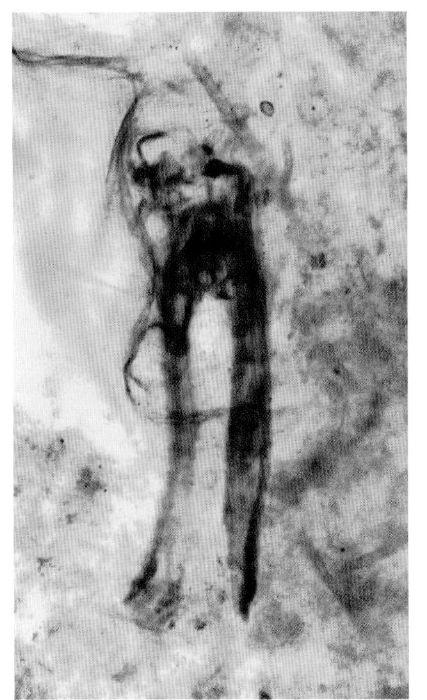

➢ The controversial fossilized mandibles of *Rhyniognatha hirsti*, a possible early insect.

# STUDYING INSECTS

*Our understanding of what insects are made of and how they live their lives is constantly growing, thanks to the efforts of scientists working in labs and wild places all around the world.*

The study of insects is called entomology. However, entomology as a field of study draws upon many other disciplines too, including genetics, cell biology, comparative anatomy, ecology, biogeography, and biochemistry. To understand fully how any given species functions, we need to look not only at its anatomy and cellular structure and its physiological processes during its life cycle, but at how, as a living insect, it uses its environment and how it fits into ecosystems.

Insects lend themselves to study in the lab in many ways, as they are mostly small, easy to keep, and breed rapidly, so it is easy to obtain specimens for tissue samples, and easy to study living processes and behaviors under controlled conditions. Studying the activities of insects in the wild is a very different matter. Even tracking one individual's movements over more than a few seconds can be impossible, although satellite tagging technology is evolving so rapidly that it may not be long before we can fit a flying insect with a tag light enough to allow it to fly unimpeded.

Ecology as a field of research has always drawn on data from "citizen science" projects, through the observations made by keen amateurs, and here is where nonspecialists have much to contribute to furthering our knowledge of insect behavior and distribution. Most of us will not be able to get our hands on an electron microscope or a DNA sequencing machine, but we can all head outside with a notebook and document what we see. In the UK, for example, volunteers

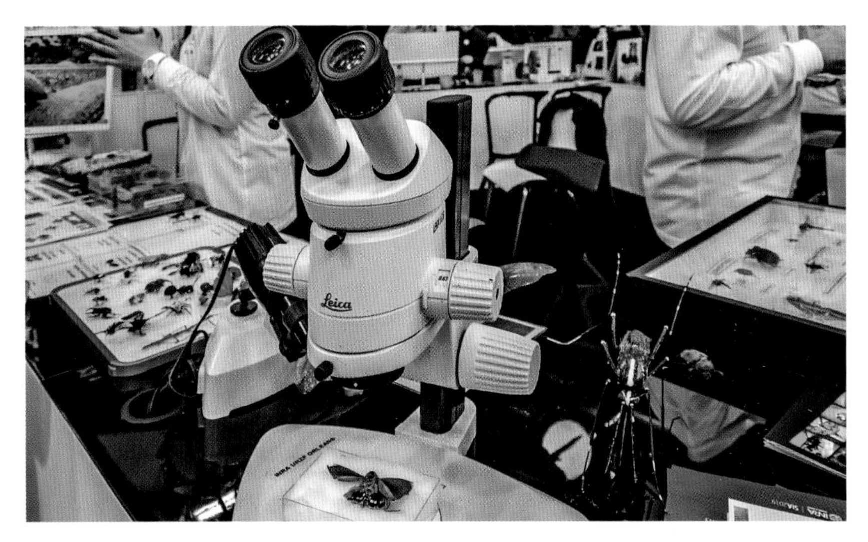


Ⓐ Some aspects of insect study involve lab work and the use of some very specialized equipment.

Ⓐ Any garden or outside space will offer plenty of interest for the amateur entomologist.

Ⓒ Pond-dipping is a great way to study the aquatic larvae of insects such as mayflies, damselflies, and diving beetles.

take part in one of the North American Butterfly Association's Butterfly Counts. This scheme, begun in 1975, now has 450 count sites, each a circle with a 15-mile (24 km) diameter, and compiles all butterfly observations from any point in the site over a one-day period. The resultant data has provided many insights into changes in butterfly abundance and distribution. Many other such schemes exist for other insect groups in other parts of the world.

## Team up

To further your own knowledge of insects and contribute to the body of work on insect biology and ecology, a good first step is to find your local entomology group. Most such groups are active on social media, and you may also find local groups that specialize in particular types of insects, as well as conservation organizations and local groups that work toward the well-being of all kinds of wildlife.

# THE FIRST FLIERS

**By the start of the Carboniferous Period, 359 million years ago, insects were widespread in many terrestrial habitats. However, only when flight evolved did they become a dominant force.**

Three sets of living animals (and one extinct) have evolved true flight—sustained, self-powered, and controlled travel through the air. The groups are insects, birds, bats, and the extinct pterosaurs. Insects were the first by almost 200 million years—and perhaps much longer. In 2015, a study examining the mutation rate of modern insect DNA and other genetic material delved into insects' deep evolutionary history. It concluded that they first arose nearly 500 million years ago, as the first plants were taking root on land, with winged insects appearing by 400 million years ago.

Traditionally, the whole insect group is divided into wingless apterygotes and winged pterygotes. It is commonly assumed that the apterygotes came first, although they may have had several evolutionary origins. The winged insects vastly outdo them today in variety and numbers and are believed to have had a single ancestral origin. This means that they form an evolutionary unit known as a clade.

### WAYS TO FLY

In terms of anatomy, there are several ways that insects fly, as explained in more detail on later pages. Probably first to evolve was direct flight, as shown today by the mayflies, (Ephemeroptera) and the dragonflies and damselflies (Odonata). The muscles that work the wings are attached between the inside base of the strong, box-like thorax (the middle of an insect's three body sections) and the bases of the wings themselves, inside the thoracic wall. In other groups, indirect flight involves muscles attaching to the "roof" of the thorax that abruptly flex the whole thoracic casing. The thorax "snaps" to and fro

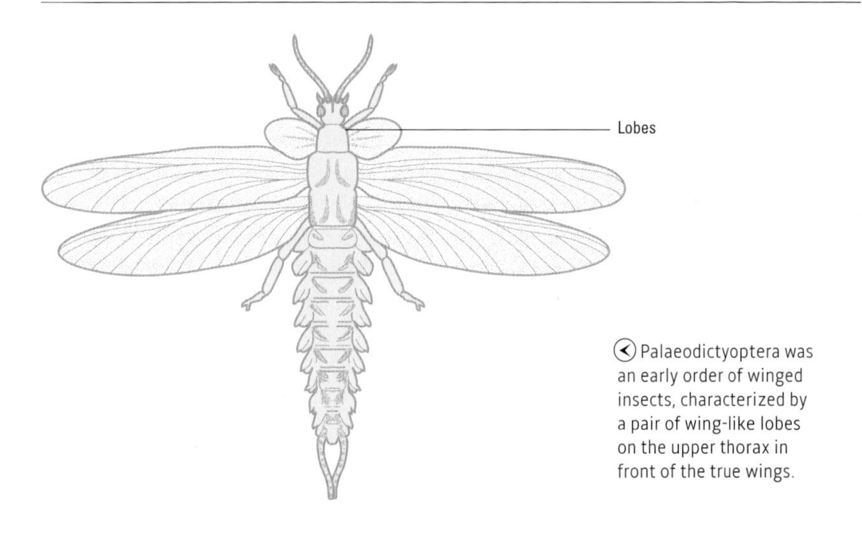


Palaeodictyoptera was an early order of winged insects, characterized by a pair of wing-like lobes on the upper thorax in front of the true wings.

between its two stable states (relaxed and flexed) to flick the wings up and then down.

**ORIGINS OF INSECT FLIGHT**

Insects' two pairs of wings may have derived from flat protrusions on what were once early limbs or protuberances from the thoracic wall. At first these possibly came in useful for emergency parachuting, perhaps from taller plants to escape predators. Gradually the walking limb part of the design faded and the flanges enlarged, permitting more controlled gliding and eventually fully flapping flight.

By the beginning of the Permian Period, 299 million years ago, more than ten groups of flying insects had evolved, including mayflies, dragonflies, grasshoppers, cockroaches, and the palaeodictyopterids—a once extremely numerous and diverse group that died out by the end of the Permian.

Ⓐ True flies, such as mosquitoes, have one pair of wings, the other being reduced to knob-like halteres that help maintain stability in flight.

Ⓥ Mayflies have existed on Earth for nearly 300 million years—nearly twice as long as mammals.

# INSECT EVOLUTIONARY TREE

**Almost all adult insects have six legs. However, other arthropods also have six limbs or similar appendages, but they are not insects. With the insects (Insecta), they are part of the larger group Hexapoda ("six legs/feet").**

As well as insects, the hexapods include three other, far less numerous and less diverse groups. The collembolans, or springtails, leap great distances for their size using a flick-action "tail." The proturans are tiny soil-dwellers nicknamed "coneheads." The diplurans are known as two-pronged bristletails, referring to the two long extensions, cerci, at their rear ends. These three groups of wingless arthropods were once considered to be members of Insecta, but are now separated within Hexapoda.

The first true insects were wingless and, as mentioned previously, are represented today by two small subgroups, Archaeognatha (rock and jumping bristletails) and Zygentoma (silverfish and firebrats). Around 300–350 million years ago, during the Carboniferous Period, a surge of evolution saw several of the main insect subgroups established, including mayflies, dragonflies, and cockroaches.

### NEW OPPORTUNITIES

When new sources of food, habitats, and other opportunities become available, evolution tends to progress rapidly in bursts called adaptive radiation. This happened several times for insects. Around 270–290 million years ago, the largest of all insect subgroups arose, the beetles and weevils, Coleoptera. Another subgroup dating from this time was the scorpionflies, Mecoptera. These mecopterans

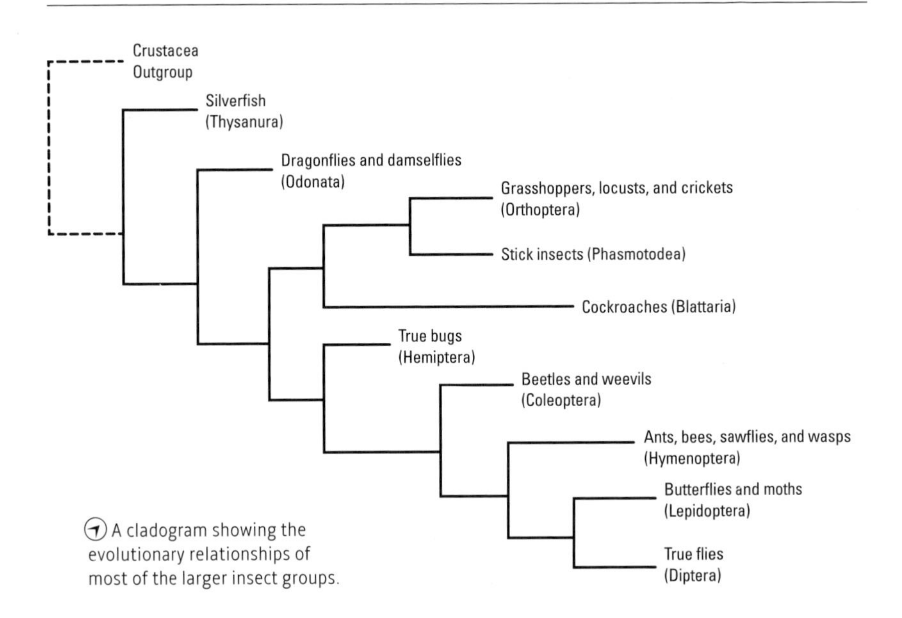


A cladogram showing the evolutionary relationships of most of the larger insect groups.

or their close relations had probably given rise to the two-winged or "true" flies, Diptera, by 240–250 million years ago.

## MORE INSECTS EVOLVE

Other lineages from mecopteran-like ancestors include the caddisflies, Trichoptera, appearing some 230 million years ago. Their close cousins the moths and (later) butterflies, Lepidoptera, arose around 180–200 million years ago. The fleas, Siphonaptera, are also part of this Mecoptera-based insect radiation, with early flea-like forms known by 150 million years ago.

Another major insect subgroup is Hymenoptera—the sawflies, wasps, bees, and ants. These lineages appeared alongside early dinosaurs in the Triassic Period, more than 220 million years ago. By the beginning of the Cretaceous Period, 145 million years ago, nearly all of today's main insect subgroups were established.

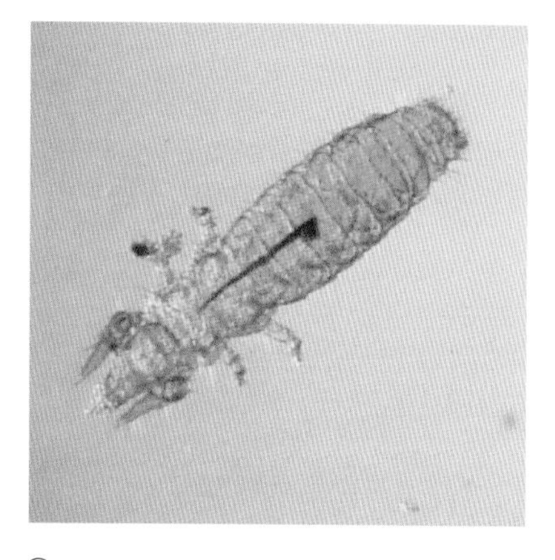

Ⓐ A "conehead," belonging to the order Protura. These arthropods are six-legged and are closely related to the insects.

Ⓥ Springtails and the other non-insect hexapods differ from true insects in several ways, including having internal rather than external mouthparts.

# A PLACE ON THE TAXONOMIC TREE

*How to classify life on Earth has preoccupied biologists for centuries. Through fossil evidence, comparative anatomy, biogeographical data, and genetics, we are building an accurate family tree for all living things.*

The study of taxonomy concerns working out the evolutionary relationships of different animals. In the case of our own species, we have good evidence that the living animals most closely related to us are the chimpanzees, and that we and they both descend from an ancestral species that lived some six or seven million years ago.

Insects and other arthropods form a major subdivision of the animal kingdom—the group Arthropoda. Their closest living relatives are other invertebrate groups including the tardigrades or "water bears," the nematode worms, and the velvet worms. Their shared ancestor, which lived more than 530 million years ago, would have been a marine animal with a soft, segmented, modular body. The earliest arthropod fossils date back about 510 million years.

Insects evolved from the first arthropods to begin to live on land around 419 million years ago. Traditionally, we classified insects and crustaceans as two distinct groups of arthropods, which went their separate evolutionary ways some 400 million years ago, evolving their distinct traits after this split. However, studies of DNA sequences in both groups have shown this is not the case, and that insects in fact evolved from one particular lineage of crustaceans. Modern insect and crustacean classification places them together in a group called Pancrustacea.

## Studying taxonomy

The fine detail of anatomy is important in working out how closely related different animals are. Without this level of study, we might look just at superficial traits and make mistakes, such as thinking dolphins are fish, or bats are birds. Accurate dating of fossils is also important, showing when particular traits first appeared in a species' lineage. Looking at biogeography can show us where certain groups originated and their patterns of spread around the world.

However, the study of DNA gives us our most powerful tool yet for studying relationships. Through comparing gene sequences, and applying our knowledge of gene mutation rates, we can work out how closely related two animals are, and how long ago their last shared ancestor had lived. This relatively new line of research is rapidly producing fresh insights into insect classification.

The scientific names we give to insects (denoting their species, genera, and families) reflect our current understanding of their relationships with other species (see page 22 for more about this). New genetic research has led to the names of several insects being changed to reflect new knowledge, which means that field guides need to be regularly updated with any changes.

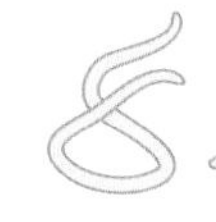

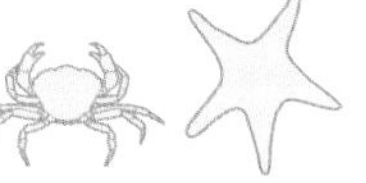
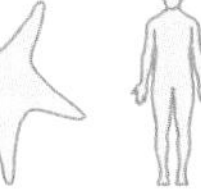


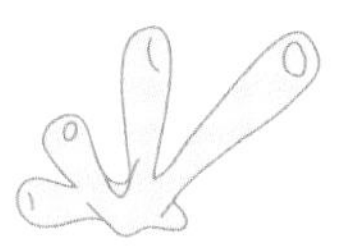


Ⓐ A simplified cladogram of the principal animal phyla, including appearances of key traits such as bilateral symmetry, showing where arthropods fit in alongside their closest relatives—the annelid or segmented worms. The key difference between protostomes and deuterostomes is in their embryology—in the former, the point where the early embryo first folds over is destined to become the mouth, but in deuterostomes it becomes the anus.

# INSECT CLASSIFICATION

*The main groupings of insects are known as orders, and there are about 30 different insect orders in the world. Each order is subdivided into several further categories.*

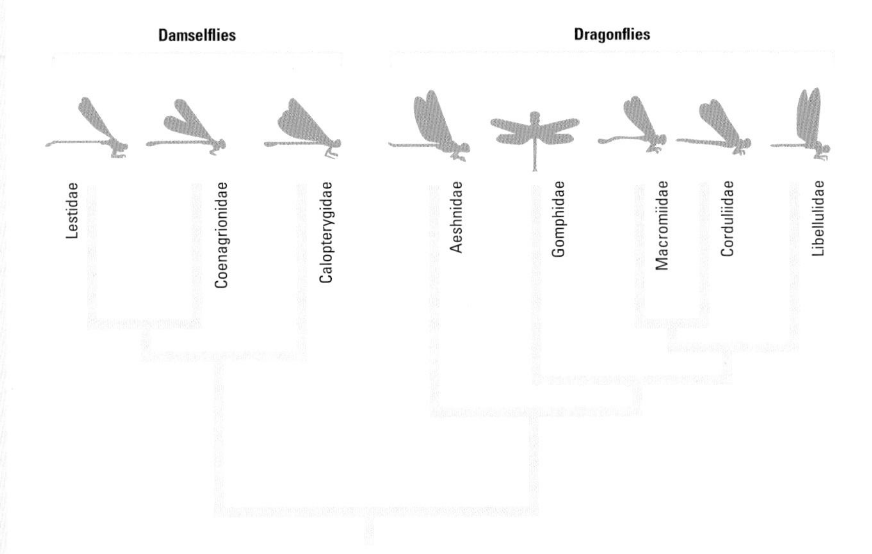


Ⓐ The relationships within the insect order Odonata. Its two basic groups—the dragonflies and the damselflies—have a single common ancestor, and each branches off into several distinct families, some of which are shown above.

Classification in biology, sometimes referred to as taxonomy, is a difficult discipline. Real nature is not arranged in as orderly a manner as we would like. Arguments between taxonomists are frequent, and compromises have to be made to come up with a workable system.

Traditionally, an order is broken down into one or more families, the members of each family sharing particular traits. For example, the dragonfly order Odonata contains about 30 families, including Aeshnidae (the hawker dragonflies—large, parallel-edged bodies, fast-flying), Calopterygidae (the broad-winged damselflies, large with colored wing patches in the males), and Lestidae (the spread-winged damselflies—often metallic-colored, rest with wings held away from the body). Each family contains one or several genera, which share a narrower range of traits. For example, Aeshnidae includes the genera *Anax*, *Aeshna*, and *Brachytron*, which all show distinctly different head shapes in their larval form. Finally, within each genus there is one or more individual species.

This naming system sets out to group insects according to how closely they are related to other species and other groups—how long ago in evolutionary history they shared a common ancestor. However, evolution is a slow and gradual process, better visualized as a branching tree rather than a filing cabinet. We can introduce intermediate ranks for groupings that do not fit—suborder,

# EXOSKELETON

**The adult insect's body is supported and protected by an exoskeleton, a more or less rigid outer shell that offers protection from the elements and helps prevent water loss.**

In vertebrates, the body's support system comes from the bones within—the endoskeleton. An insect's exoskeleton provides support from the outside, at the expense of some bodily flexibility. However, although the exoskeleton's constituent parts are rigid, the joints between the segments allow the insect to move its body sections—in some cases very freely. It also fractures at certain joints in immature insects prior to being shed (molted) and this process enables the insect to grow.

The outer covering, or integument, of an insect is known as a cuticle. It is an efficient multipurpose casing for the body, giving it the necessary stiffness to resist damage from encounters with environmental hazards, and preventing dehydration. It typically has two layers—an outer waxy covering (the epicuticle) that provides waterproofing, and the inner layer (procuticle). The procuticle is made from a soft, flexible, protein-based connective tissue called chitin. Underneath the cuticle is the epidermis, a single layer of cells that produces the proteins that form the cuticle.

In some insects (especially those that live in damp environments), the procuticle remains single-layered and soft over part or all of the body. However, in most adult insects at least part of the procuticle develops a hard outer layer (exocuticle) of more rigid proteins. The development of the exocuticle is known as sclerotization. Body parts with particularly well-developed sclerotization include the "armor" of some beetles, and the biting mouthparts of dragonflies. Most insects have a well-sclerotized head capsule, enclosing all of the segments of the head. This capsule is shed as a single piece when an insect larva molts.

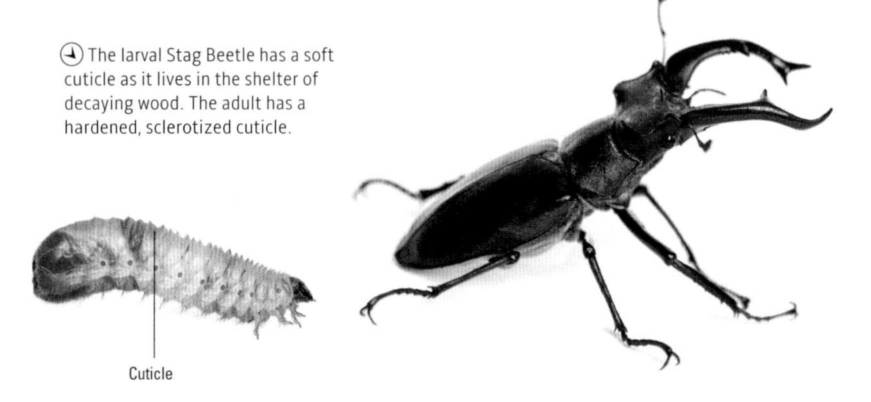


The larval Stag Beetle has a soft cuticle as it lives in the shelter of decaying wood. The adult has a hardened, sclerotized cuticle.

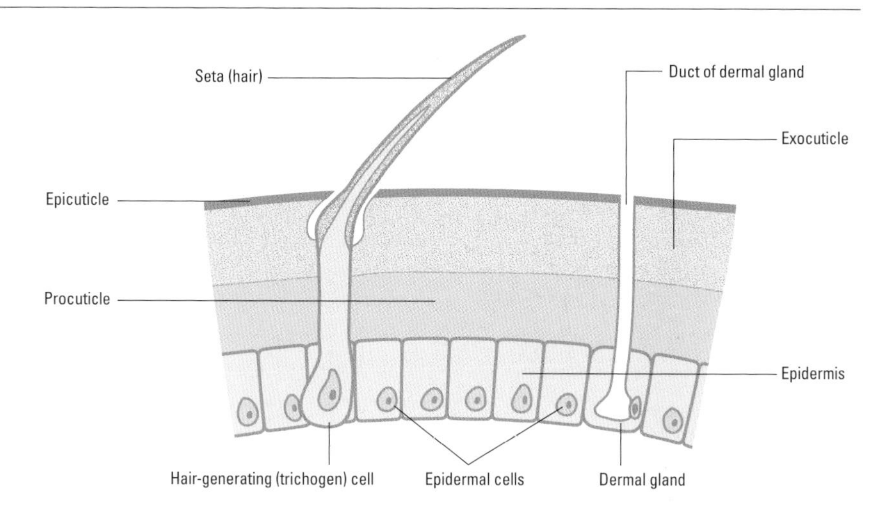


Ⓐ The layers of an insect's outer covering. The exocuticle, or outer part of the procuticle, is not present in all insects.

**HAIR AND SCALES**

Many insects have hairy or even furry bodies. The hairs or setae each form from one individual specialized cell (a trichogen cell) within the epidermis—the hair grows outward through the cuticle. The function of setae varies between species and includes protection or camouflage, heat conservation, and storage facilities (the pollen baskets of bees are formed from clumps of setae). They can even provide self-defense (the setae of some moth caterpillars contain a toxin that irritates the skin or mouth of any would-be predator).

Caddisflies, butterflies, and moths have modified setae, which are expanded into wide, plate-like scales. Layers of scales cover the bodies and most noticeably the wings of moths and butterflies—they provide protection, play a role in heat absorption, and their microstructure interferes with light waves to produce the diversity of iridescent color found in many members of this group. The wing scales often get dislodged, but the membrane below is still flight-capable.

Ⓥ Most insects have quite plentiful hairs, or setae, on at least some parts of their bodies.

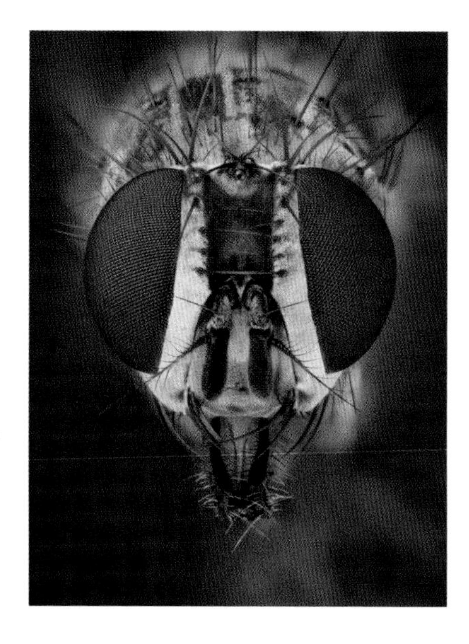

# THE THREE BODY SECTIONS

**One of the first things we learn about insects' bodies, besides the fact that they have six legs, is that their bodies are comprised of three distinct sections—the head, the thorax, and the abdomen.**

A close look at a wasp or an ant makes clear that its head, thorax, and abdomen are very distinct, the attachments between the three parts being marked by a pronounced narrowing of the body. In other insect groups the distinctions are less obvious, but still apparent on examination.

Like almost all freely moving animals, insects have their primary sensory and eating equipment at their front ends, to gather information about whatever they are approaching (and then, if it is food, to consume it). The head usually bears two obvious compound eyes and a pair of antennae. There will also be a set of mouthparts, which vary greatly across different groups of insects in the way they are shaped and how they work. In a few cases, adult insects have no functional mouthparts and do not feed at all during their short lives. The antennae and mouthparts are all modified versions of the uniramous appendages that grow in pairs from the segments of the head, and are segmented, like the legs.

The three segments that form an insect's thorax, from front to back, are the prothorax, the mesothorax, and the metathorax. Each bears a pair of legs (though one or more pairs may be rudimentary in certain species), used for walking, running, swimming, jumping, or just perching. The thorax is also the attachment point for the wings in flying insects, with the front pair attached to the mesothorax and the hind pair to the metathorax. The thoracic segments also contain musculature to drive the wings.

The segments at the thoracic end of the abdomen lack appendages, but the segmented structure is usually quite evident, and often even accentuated by contrastingly

Wasps and related insects show very distinct body sections, but all other insects' bodies are similarly divided into head, thorax and abdomen.

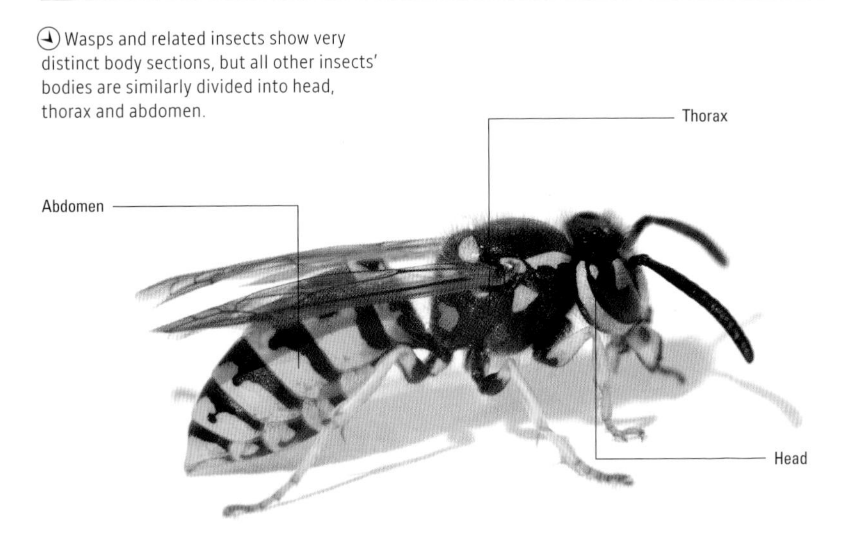

Ⓐ Earwigs have prominent cerci—the paired appendages on the last segment of the abdomen.

Ⓐ Damselflies have very long abdomens, and boxlike thoraxes that bear their four large wings and long legs.

Ⓐ Stoneflies' wings lie flat along their abdomens when they are at rest.

Ⓐ Mayflies lack functional mouthparts in adulthood, so they have small and relatively simple heads.

often even accentuated by contrastingly colored markings. Some insects have 11 or 12 abdominal segments, but it varies—more recently evolved groups have fewer. The insect's reproductive organs, which may have visible external parts, are found near the abdomen tip, while the last segment has a pair of appendages (cerci). These are prominent in some groups (as in the "forceps" of ear wigs) and often have a reproductive function—for example, male dragonflies use their cerci (or "claspers") to grip the female prior to mating.

**JOINING UP**

The neck, or cervix, that links the head to the thorax is membranous, with muscles to control head movement. Many insects can move their heads freely. The junction between thorax and abdomen is less flexible, as the first abdominal segment (the propodeum) is fused to the metathorax. The slim "waist" of some insects, such as wasps, is formed by a particularly narrow propodeum, giving these insects more range of movement between the thorax and the abdomen.

# SEGMENTATION AND APPENDAGES

**In all arthropods, the body is made up of repeated segments, joined together front to back. That is very apparent in a caterpillar, but much less so in an adult butterfly.**

The standard, most primitive arthropod body is made up of a repeated series of near identical body segments, each of which bears a pair of jointed appendages. In some arthropods, the appendages are biramous (split into two branches), and in most such cases one of the branches of a biramous appendage functions as a leg and the other as a gill. This structure can be seen in crustaceans, such as crayfish. However, in insects the appendages do not have the second, gill branch, and are classed as uniramous (single-branched).

Very few insect species show, when adult, a continuous body structure made of repeated, similar-looking segments. Instead, the body segments (if apparent at all) are arranged in distinct groups, and their shape is modified according to their function and position in the body. Even the insect's head was originally formed of five to seven distinct segments, although these are not apparent on an intact adult insect as they are fused into a continuous capsule.

The abdominal segments do not have appendages. Those of the thorax and head do bear appendages but these are uniramous. The appendages of the thorax function as legs, while those on the head form feeding and sensory apparatus.

The insect's leg, like its body, has a segmented structure, with moveable joints between some of the segments. Its main parts are the coxa (the first segment, where it joins the body—typically very short), the trochanter (second segment, also small and functioning like a hip joint in vertebrates), the femur (the third and usually largest and thickest segment), the tibia (the fourth and often longest segment, articulating with the femur in a flexible, knee-like joint), and the tarsus (formed of several small segments, or

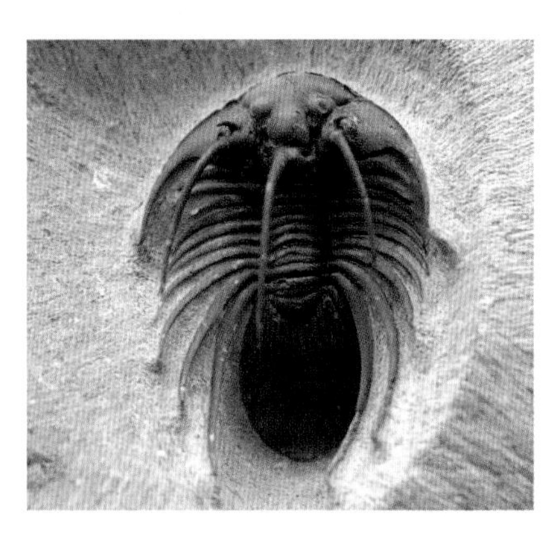

Ⓐ The trilobite genus *Kolihapeltis* was notable for the long spines that project from the head segments.

Ⓐ *Stylonurus* was a genus of chelicerates—the arthropod group that contains the spiders and scorpions. The body and leg segmentation was relatively simple.

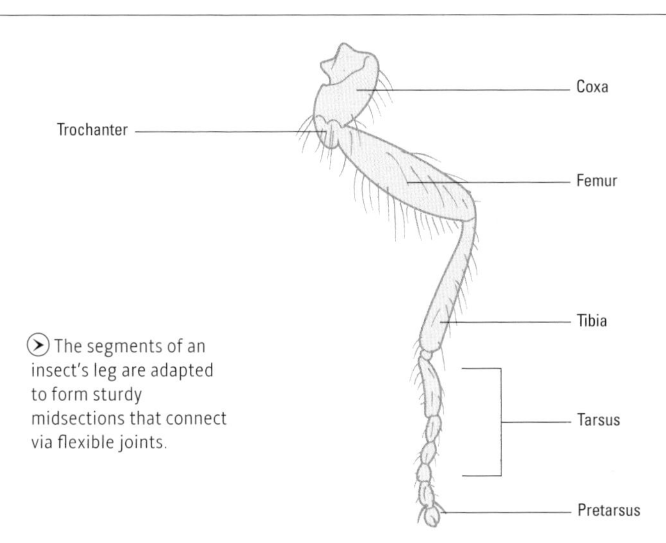


The segments of an insect's leg are adapted to form sturdy midsections that connect via flexible joints.

tarsi, giving some flexibility) and forming the last joint. The last tarsal segment, the pretarsus, supports the claws.

## INTERNAL ORDER

The body segments also show some repeated structural features with regard to their internal anatomy—this can be seen with aspects of the nervous system and circulatory and respiratory systems, for example. The insect body plan is sometimes described as modular because of that repeated arrangement, with each segment having some degree of autonomy from its neighbors.

Many caterpillars show obvious body segmentation, and the three pairs of true legs at the front of the body are jointed.

Adult insects of the more recently evolved groups, such as this chrysid wasp, lack a uniformly segmented appearance. The head and thorax in particular look like single structures.

# LIMB STRUCTURE

**Insects use their legs for much more than locomotion, including such behaviors as gripping prey or a mate, grooming, and performing social displays. Whatever the legs' function, though, all share the same basic plan.**

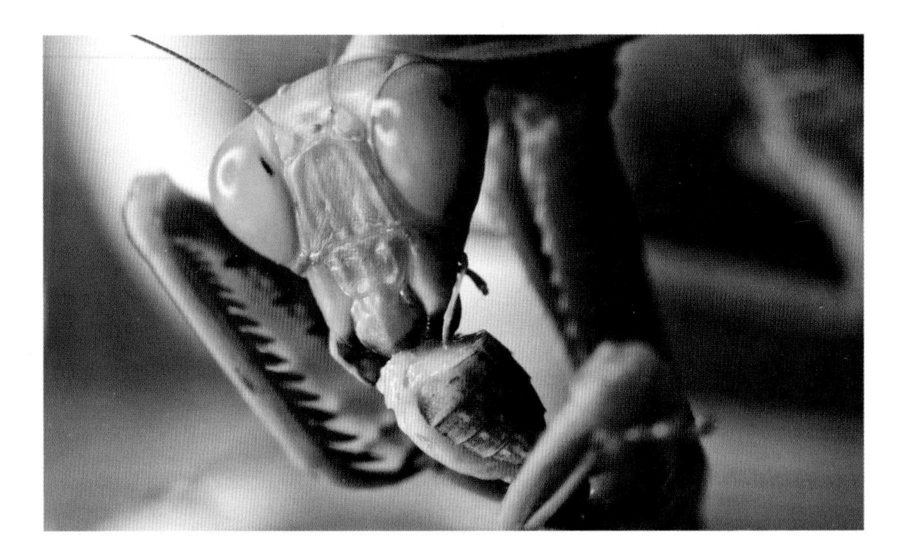

Ⓐ Mantises use their long, barbed forelimbs to catch and grip prey.

As we have seen, insects' legs are segmented and have five main sections, each formed by a single segment except the outermost one (the tarsus), which is formed from a "stack" of several small tarsi. The five sections are attached to each other through flexible hinge joints. The integument of the limbs is the same as for the body, comprising a cuticle that overlies the epidermis. Legs often have plentiful setae. The dense long hairs on the tarsi of diving beetles and bugs function as swimming "flippers," while the front legs of dragonflies and damselflies have rows of long, stiff setae that form a sort of net to help capture prey on the wing.

The last section of the tarsus, the pretarsus, usually has one or two small hooked claws, which help the insect hang on to the substrate as it walks or rests. Other structures that may be found on this segment include hairy adhesive pads or lobes, which provide grip on flat surfaces.

In addition to the main sections, legs may have additional projections. The forelimbs of praying mantises, for example, have sharp spines and serrations for piercing prey. In burrowing or burying insects, such as mole crickets or scarab beetles, the tibia or tarsus has flat, hard projections for digging into earth.

## SHAPED FOR LIFE

The relative size and shape of insects' leg sections can provide clues to the way they live. Long but sturdy legs indicate a fast runner, while more delicate legs suggest an insect that perches or climbs. Jumping insects

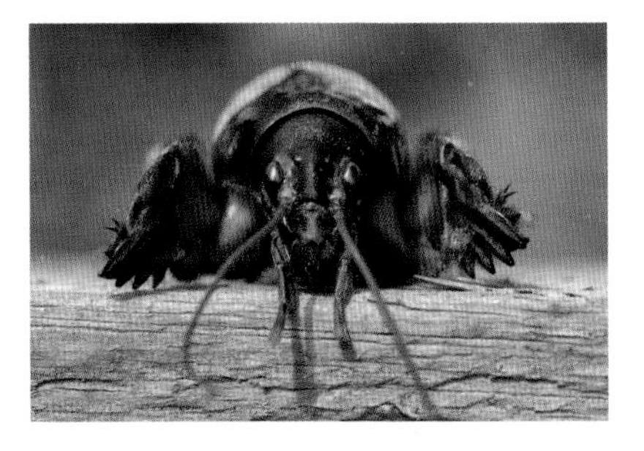

Ⓐ The hefty, clawed front legs of mole crickets are used for digging in soft earth.

Ⓐ Millipedes and other myriapods are arthropods, like insects, but have a simpler body plan and far more pairs of legs.

Ⓐ Most beetles can fly, but have strong flexible legs and travel more on foot than in the air.

Ⓐ Worker ants lack wings, but their long legs enable them to climb easily and run quickly.

usually have hind legs that are much longer than the front or middle pair, and show disproportionately large femurs to power their leaps. Digging insects have thickened front legs, and predators often also have larger and longer front legs. Swimmers may have one or more pairs of specialized, fin- or oar-like swimming legs.

Ⓐ Phasmids use their long, delicate legs to climb through plant foliage.

# THE WINGS AND ELYTRA

**Most insects are winged in their adult form and can fly. Insect wings are often breathtakingly beautiful in their structure, perhaps even more so when viewed under a microscope.**

The power of flight has allowed insects to spread and thrive throughout even the most unpromising habitats on Earth. Some spend almost their whole adult lives on the wing and fly with extraordinary efficiency and agility, while for others the wings are weak and are an "emergency" option to escape a sudden threat. Despite this diversity, all winged insects are thought to share a single common ancestor, and their wings show many commonalities.

Most insect groups have two pairs of transparent wings, formed from a fine membrane and supported by a branching network of veins, which may be nearly invisible or dark and prominent. There is typically a particularly long and sturdy vein (the costa) at (or almost at) the leading edge of the wing—the insect can usually fly even with considerable wing damage as long as the costa is intact. The exact pattern of wing venation is important when it comes to telling apart some confusingly similar insect families.

In some groups, including the bees, wasps, and mayflies, the hind wings are considerably smaller than the forewings. Among the grasshoppers the forewings are long and narrow, with a thick and leathery

Ladybugs and other beetles keep their hind wings folded up under their elytra. They can only fly once the elytra are opened and the wings unfurled.

Ⓐ The dark smudges on the forewings of the male Gatekeeper Butterfly are formed by androconia—special scent-producing scales.

Ⓐ Butterflies use their wings to regulate their temperature, opening them to warm up and closing them to prevent chilling or overheating.

texture. When the insect is resting the forewings protect the hind wings, which are membranous and shorter but very broad. In the true flies (Diptera), the hind wings are reduced to small club-like structures (halteres).

Caddisflies have dense setae on their wings, and, as we have seen, those of butterflies and moths have layers of overlapping scales. These include, in males of some species, specialized scent-releasing scales (androconia)—the insect wafts this scent toward a female during attempts to mate. Some membranous-winged insects have patches of pigment on their wings, forming patterns that may be shown off in wing-flicking courtship displays.

#### ELYTRA

The elytra, or wing cases, of beetles are modified forewings. They are heavily sclerotized, thick, and strong, protecting the hind wings, which are folded beneath, as well as the abdomen. When the insect takes flight, its elytra lift and part to allow the hind wings to unfold. The elytra are held in this raised position during flight. Some true bugs have partially sclerotized forewings, known as hemelytra.

The elytra do generate some lift in flight, but their primary purpose is protection when the beetle is on the ground—in the air, beetles are much clumsier than most four-winged insects, and their flight is less energy-efficient. However, the elytra's sturdiness enable beetles to dig tunnels, scramble through thorny vegetation and generally survive conditions which would quickly tear the exposed wings of other insects to shreds. The elytra are also a canvas for color—dull tones that blend in, bright warning patterns, or vivid iridescence—which has various possible functions including both camouflage and signaling.

# UNUSUAL BODIES

**Although the adult forms in all insect groups have three-sectioned, six-legged, exoskeleton-supported bodies, within these constraints, there are some remarkable variations in body form.**

Some of the most bizarre-looking insects in the world are the rhinoceros beetles (family Scarabaeidae), males of which have large, rigid sclerotized horns on their heads and thoraxes, which they use in battle when competing to access females. In the enormous Hercules Beetle (*Dynastes hercules*), found in South and Central America, the thoracic horn may be longer than the rest of the body and extends even beyond the insect's sizable mandibles. Males of the Giraffe Weevil (*Trachelophorus giraffa*) of Madagascar also have exaggerated anatomy for fighting purposes—in this case, a long, slim, giraffe-like "neck" formed by an elongated head and thorax.

Camouflage is another driver for unusual physical modifications to evolve, such as the flattened extensions of cuticle on the legs and

Ⓐ Butterflies of the genus *Kallima* replicate the appearance of dead leaves when resting with the wings closed.

Ⓥ The leaf insects (family Phylliidae) have been shaped by evolution into stunning leaf mimics.

Male rhinoceros beetles (subfamily Dynastinae) use their "horns" to battle with other males and to dig.

bodies of leaf insects (family Phylliidae) and the Orchid Mantis (*Hymenopus coronatus*), a southern Asian species and one of several mantises with a petal-like appearance. One of the longest insects in the world, the 22-inch (56cm) phasmid *Phobaeticus chani* of Borneo is a twig mimic with very slender and elongated body and legs.

The rove beetles (family Staphylinidae) have undersized elytra, leaving a considerable length of segmented abdomen fully exposed. This gives these ground-dwelling beetles more bodily flexibility, enabling them to wiggle into small spaces. Most species have full-size wings and can fly, though in some species the elytra are fused.

## EVERSION THERAPY

Some insects have body parts that are normally held within the body but can be turned inside out (everted) to protrude outward. These are usually involved in the release of volatile chemicals into the atmosphere—pheromones to attract a mate, or repellent substances to ward off predators. One of the most spectacular examples is provided by an Asian and Australasian moth called *Creatonotos gangis*. Males of this species have a pair of very large, forked organs called coremata, which are covered with long, fine hairs. When everted from the abdomen's tip, the coremata may be much longer than the rest of the moth's body—their size depends on the foods the moth ate as a larva. The hairs of the coremata release a potent pheromone to attract females.

# INSECT COLORS

*Many insects, from beetles to bees and dragonflies to deerflies, have dazzling coloration. Its purpose may be to entice, disguise, or warn.*

The white light of the Sun contains all of the different wavelengths of light that we see as color—we can see the full rainbow of tones when we shine white light through a glass prism, causing refraction (bending and splitting of the light beam). Different shades are produced by the properties of the material that the light strikes.

In nature, color is formed in two different ways. The first is through pigmentation—molecules within the organism's cells that absorb light of certain wavelengths and reflect back others, which we see as visible color. The second is structural coloration, where the body has some physical feature that bends or scatters white light into different wavelengths. Here, the viewing angle can affect the color tone that we see, and the color is often brightly shining as well as constantly changing—this is known as iridescence, and most commonly produces blue, green, and violet tones.

In many insects, the colors we see are a mixture of pigmentation and structural color. Iridescent butterflies produce their coloration through the shapes of the scales on their wings, and iridescent beetles have ridges on their epicuticles that bend light, but both have underlying layers of pigment cells that enhance the effect. Dragonflies of some species have dark-pigmented bodies but develop a thick waxy bloom (pruinescence) on the cuticle, which creates pale blue tones through light-scattering.

Pigmentations found in insects include melanins (responsible for dark browns and grays), carotenoids (giving red, orange, and yellow tones), and many more. Some pigments are synthesized in the body, while others are sequestered directly from food the insect eats—foods eaten as a larva may provide pigmentation for the adult body.

Many insects' colors are drab, for camouflage, but bright and even iridescent coloration can be effective camouflage in the right sort of habitat. Butterflies, with the ability to close their wings and thus completely hide the upper sides, can have the best of both worlds—a camouflaged underside for hiding and a colorful upper side to appeal to mating partners.

## Ultraviolet and red

The ultraviolet part of the light spectrum is invisible to human eyes, but many insects can see ultraviolet light. This means that some of their colors are not visible to us, but are to each other. By contrast, most if not all insects cannot see red light. Red, carotenoid-based colors are present in many insects nonetheless, often as warning coloration, for example, on the elytra of ladybugs. It is there for the benefit of predatory birds that can see red light—these insects are toxic or they taste bad, and if a bird eats one it will remember that similarly bright red insects are to be avoided in the future.

➚ Natural colors in the insect world (clockwise from top left): Paper Kite Butterfly; Monarch Butterfly; Seven-Spot Ladybug; Pale Tussock Moth Caterpillar; Buff-Tailed Bumble Bee; Rose Chafer Beetle; Banded Jewel Beetle; Scarce Chaser Dragonfly

# THE SENSES AND THE NERVOUS SYSTEM

Insects are bombarded by the same stimuli that we are, but their sensory and nervous systems work very differently. From the multifaceted structure of their eyes to the taste buds in their feet and their whole-body brains, insects experience and respond to the world in ways quite alien to us.


3.1 • Eyes

3.2 • Antennae

3.3 • Chemoreception

3.4 • Hearing, touch, and more

3.5 • Brains, ganglia, and nerves

3.6 • Insect intelligence


The wasp's large, prominent eyes and antennae provide it with a constant stream of sensory data about the world around it.

# EYES

**Insect eye anatomy is dramatically different from our own, but the size and complexity of some insects' eyes leave us in no doubt as to how well-developed their visual abilities are.**

Ⓐ The eyes of a dragonfly wrap around nearly the whole of its head for almost 360-degree vision.

Animals that fly in the open in daylight, as many adult insects do, need well-developed, fast-reacting vision to navigate their environment. This is at its most extreme in the dragonflies, which need not only to evade obstacles and larger predators, but also to pursue and capture other flying insects at high speed. These insects have immense, "wrap-around" eyes, which occupy almost the entirety of the outsides of their heads. Flower-feeding insects such as bees and hoverflies also have very large eyes.

The primary visual organs of an adult insect are compound eyes. Under the microscope, the surface of the eye is revealed to be a tight arrangement of hexagons. Each hexagon is the surface of an individual light-sensing structure—the ommatidium. A large dragonfly has about 30,000 ommatidia per eye. Ommatidia are about ten times longer than they are wide, and narrow considerably from their surface to their base.

Each ommatidium is topped with a cornea and a crystalline pseudocone, which acts as a focusing lens, directing light into the rhabdom, a long, narrow, and transparent structure at the center of the ommatidium. It contains photosensitive pigments that respond to certain wavelengths of light. The rhabdom is formed by the combined inner parts of (usually) eight specialized nerve cells—photoreceptors. When the rhabdom's pigments undergo chemical change in response to light, these cells send a nerve impulse to the brain.

The ommatidium also contains six pigment cells, which absorb light that strikes the cornea at an indirect angle. This ensures that the photoreceptors only receive light that passes through the cornea directly. Insects cannot move their eyes to follow a point of interest, but the pigment cells provide an alternative means of retaining focus.

What the compound eye "sees" is, as far as we can tell, a scene formed by an array of colored specks (including, in some cases, ultraviolet "color"), each speck contributed by an individual ommatidium. In dragonflies, there are enough specks to form a detailed picture, but in insects with fewer ommatidia the compound image has little detail.

This Ashy Mining Bee's ocelli resemble small glossy beads, on the upper part of the head in between the two large compound eyes.

## OCELLI

In addition to the compound eyes, many adult insects have secondary visual structures called ocelli, and these are the only visual organs in most insect larvae. They are noticeable on the heads of bees and wasps, for example, as a trio of small, shiny bumps in the space between the compound eyes.

The ocelli have a cornea and a layer of photoreceptive cells, but unlike those in the compound eye, these cells only sense light levels, not color. They are thought to have an important role in maintaining stability in the air in fast-flying insects.

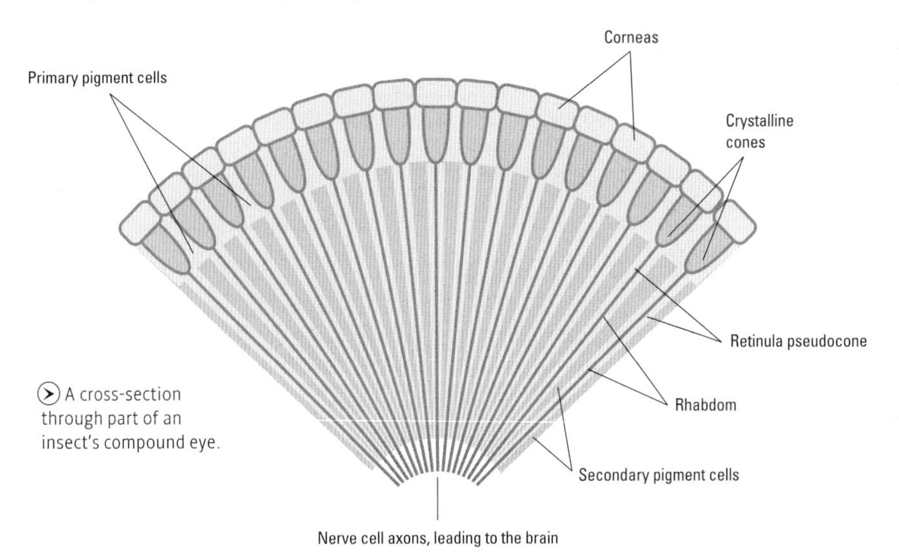


A cross-section through part of an insect's compound eye.

# ANTENNAE

**The paired antennae, or feelers, are, after the compound eyes, the most noticeable sensory structures in insects. They are highly diverse in form and also in function.**

The antennae are the appendages that develop from the second segment of the insect head. Like the legs, they have a segmented structure, which can give them great mobility. The constant antenna movement of a longhorn beetle as it walks across a flower-head shows how actively it uses these appendages to collect information about its environment.

Each antenna has three main sections. The base segment is the scape, which is small and sits within a sclerotized ring projecting from the head capsule. This connection is springy and flexible, allowing free movement of the antenna as a single unit. The second segment is the pedicel—this is usually relatively long and attached by a hinge joint to the outer section, the flagellum, which is often formed by many small segments or flagellomeres. The flagellomeres contain odor-detecting cells, and the joints between them permit this outer part of the antenna to move with considerable flexibility, allowing the insect to pinpoint the source of a scent. The pedicel contains a bundle of sensory cells (the Johnston's organ) that responds to involuntary movement of the flagellum. It can function as an organ of hearing, sensing air vibrations produced by sound, and its input can also help the insect maintain control of stability during aerial twists and turns.

Antenna shape varies from simple and straight to bulbous-ended, feather-like (plumose), serrated, or tipped with a fan of blades. The plumose antennae of certain male moths are very large, enabling them to

Ⓐ Some impressive antennae (from the top): Black Dancer Caddisfly, Gypsy Moth, Alpine Longhorn Beetle.

Ⓐ Direct contact between the antennae is a precursor to mating in many butterfly species.

Ⓒ Male fairy longhorn moths gather on sunlit vegetation and perform short display flights to attract females, their long white-tipped antennae wafting about and catching the light.

sense molecules of the female's airborne pheromones at extremely low concentrations (and to react to them with remarkable speed).

## SOCIAL SIGNALING

Some insects use their antennae to convey messages to other insects. The extraordinarily long, white-tipped antennae of fairy longhorn moths (family Adelidae), for example, dance eye-catchingly in the air as the males perform their communal courtship display dances. The presence of certain chemical compounds on ants' antennae enable them to recognize whether other individuals they encounter are from their own nest or not.

# CHEMORECEPTION

**The senses we experience as smell and taste are both classed as chemoreception, as they involve sensing particular chemical compounds through direct contact of that compound's molecules with our sensory cells.**

Although they do not have nostrils or tongues laden with taste buds, most insects are still highly sensitive to smells and tastes. Without these senses, they would have difficulty finding the right food, detecting mating partners, and locating suitable places to take shelter either for hibernation or to lay their eggs.

The organs of chemoreception are called sensilla. These small structures are located on the outside of the cuticle, and consist of one or a few receptor cells plus a few accessory cells, which moisten and protect the receptor cells, as well as the endings of the nerve cells that communicate with the receptor cells. When a molecule of a particular compound makes contact with a receptor cell, it binds to a certain part of that cell's membrane, causing a chemical change within the cell. This in turn stimulates the nerve cell that is in contact with the receptor cell, sending a nerve impulse toward the brain.

The antennae are the key organs of chemoreception, their sensilla being sensitive to smell (picking up airborne molecules of volatile compounds) and in some cases also taste (picking up the same kinds of molecules through direct contact with their source). However, most insects primarily taste through their mouthparts. Moths, butterflies, and true flies can also taste through sensilla in their tarsi (the outermost part of their legs), and some insects have taste receptors on their ovipositors (egg-laying tubes).

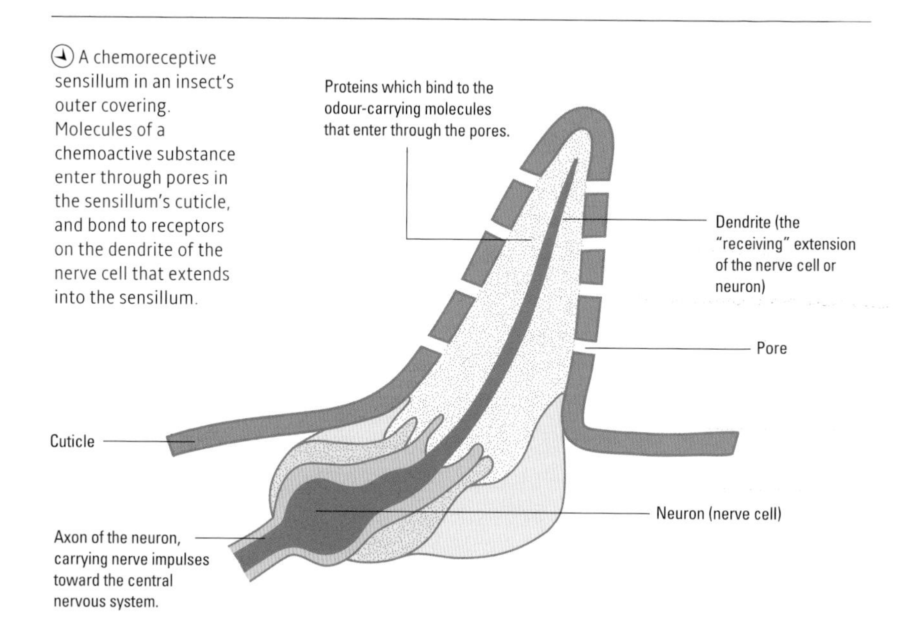


A chemoreceptive sensillum in an insect's outer covering. Molecules of a chemoactive substance enter through pores in the sensillum's cuticle, and bond to receptors on the dendrite of the nerve cell that extends into the sensillum.

A female Monarch Butterfly uses chemoreceptors on her feet to assess the suitability of a leaf before she lays her eggs on it.

Females with eggs to lay use the taste sensors in their feet or ovipositors to locate suitable leaves or other food sources for their future larvae. Insect larvae also have a well-developed sense of taste—finding food is a much more dominant driver in their lives than it is for adults, and in herbivorous insect larvae in particular, the range of foods that their bodies can process may be very narrow.

**FILTRATION**

Because there are so many volatile chemical molecules around in the air and on all kinds of substrates, the insect needs to be able to ignore all of those that are not relevant to its interests. Its chemoreceptors are therefore highly specific to particular molecule types. However, potentially dangerous substances (such as ammonia or strong acids) seem to trigger a strong avoidance response across all species of insects, even if all their known chemoreceptors have been temporarily blocked. The pathways of this common chemical response are not yet known.

Research into insects' behavioral responses to different chemicals is important when attempting to control insect species that carry deadly diseases (for example mosquitoes in areas where malaria is common).

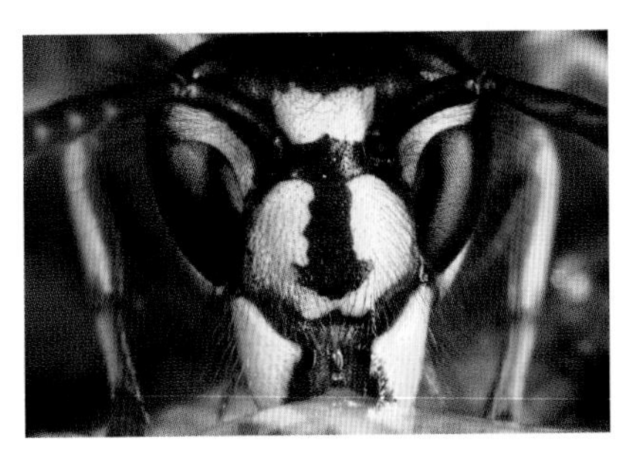

Wasps have an unerring ability to sense and home in on sweet-tasting substances.

Mantises are able to learn to avoid attacking prey that has a bitter taste.

## HEARING, TOUCH, AND MORE

**Vision and chemoreception are vitally important to most insects, but they have other sensory inputs that are valuable too, including some that seem quite alien to us.**

Ⓐ Worker ants are in constant communication with each other, through chemical and touch receptors on the antennae and other body parts.

Because insects usually seem entirely unresponsive to the sounds we make, it is easy to assume that they have no appreciable sense of hearing. However, noisy insects such as grasshoppers, crickets, and cicadas clearly do communicate through sound. Other ways that sound is important to insects are often lost on us, as they involve frequencies outside our range of hearing—for example, some moths can hear the echolocation calls of hunting bats.

The anatomical structure of hearing is the tympanal organ. It comprises a thin membrane stretched across an air-filled space, with sensory cells below. Sounds create pressure waves in the air, which vibrate the membrane. The air within is moved by this vibration, and this movement triggers the sensory cells. Tympanal organs, may be found on almost any part of the body.

Insects sense touch through sensilla, similar to those involved in chemoreception. However, the receptor cells in touch sensilla respond to contact, movement, and pressure. They can be found on all body parts. Touch sensilla often take the form of long, fine hairs, but other types include stretch receptors (which sense, for example, when it is time to stop feeding) and water-pressure receptors (which keep aquatic insects from diving too deep).

### OTHER SENSES

Some insects are also sensitive to the Earth's magnetic field, and to electrical fields. Honey bees use geomagnetic information to help them navigate directly back to their hive after

Bumblebees use their sensitivity to electrical fields to make decisions about which flowers to visit.

a long and convoluted foraging flight. A foraging bumblebee generates a small electrical field with its wingbeats, which briefly alters the electrical field of each flower that it visits. When another bumblebee comes along, it can sense this altered state, and will avoid that flower as it "knows" that the nectar has already been taken. How insects can sense magnetic and electrical fields is still the subject of debate and research.

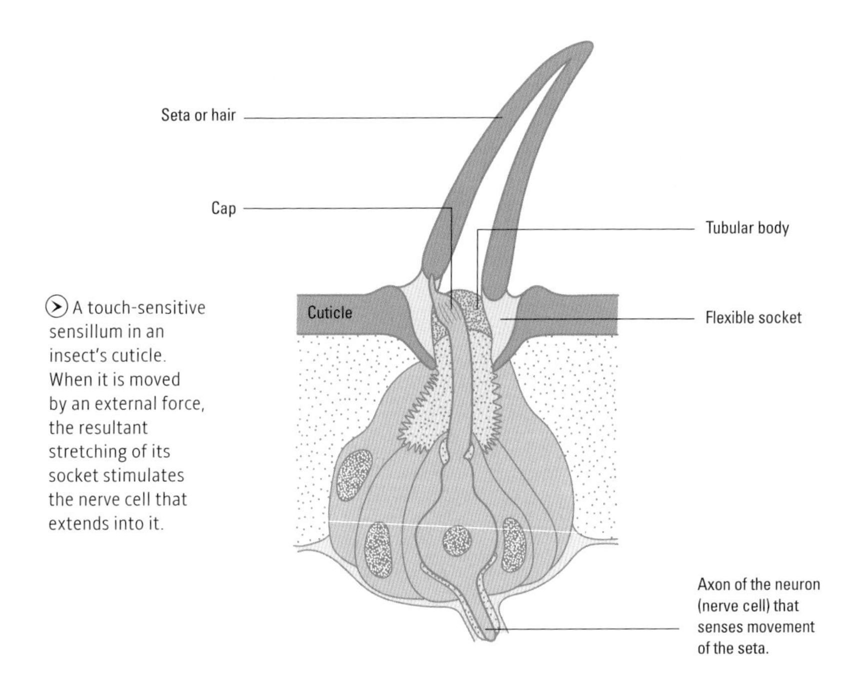


A touch-sensitive sensillum in an insect's cuticle. When it is moved by an external force, the resultant stretching of its socket stimulates the nerve cell that extends into it.

# BRAINS, GANGLIA, AND NERVES

**The input received from the insect's sense organs needs to be processed, before the insect can "decide" an appropriate behavioral response. This is the job of its central nervous system.**

The insect brain, like ours, is essentially a concentration of neurons, or nerve cells, which communicate with each other and with nerves that reach out into the rest of the body. It receives input along afferent nerves (those whose impulses head toward the central nervous system) from the sensory organs, and responds by sending out signals via efferent nerves (those whose impulses head out from the central nervous system) to the muscles, to trigger the appropriate physical movements. The efferent nerves also communicate with glands, causing them to release hormones in response to certain stimuli.

The brain is located in the back of the head, and has three distinct lobes, each region associated with the structures that arise from different head segments: the protocerebrum at the front (which is linked to the visual system), the deutocerebrum in the middle (which is linked with the antennae), and the tritocerebrum at the back (which communicates with the mouthparts).

These three regions are each formed from a pair of ganglia—a bundle of nerve endings. Though in the brain the ganglia are fused, the insect's body contains many more, separate ganglia, one pair in each body segment as a general rule (though in some cases they are fused with neighboring ganglia). In a sense, therefore, each body segment has its own miniature "brain," handling its own affairs—for example, each of the three segments of the thorax contains a pair of ganglia that control movement of that segment's pair of legs.

## NEURONS

A neuron is a cell with one or more long, filamentous offshoots (axons), which can conduct electrical impulses. An axon ends with a cluster of very fine, branching projections (dendrites), which are separated

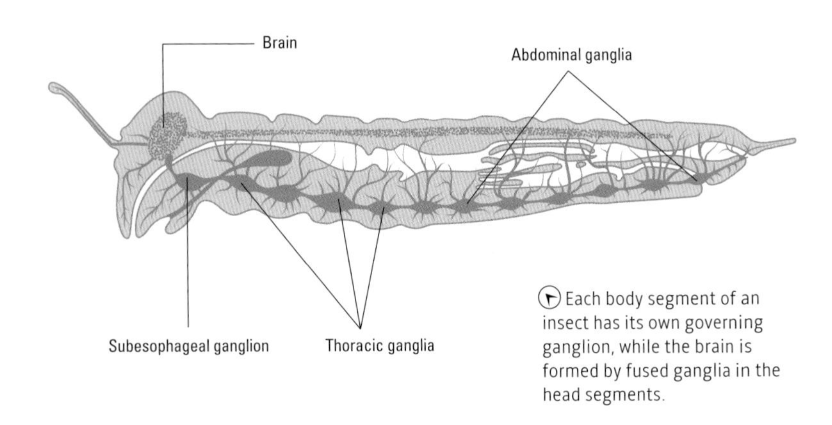


Each body segment of an insect has its own governing ganglion, while the brain is formed by fused ganglia in the head segments.

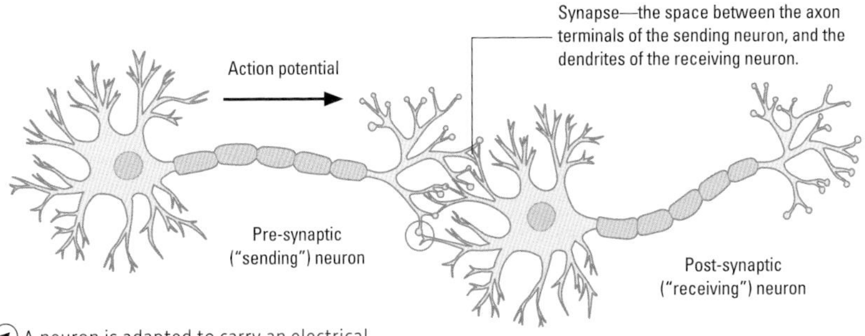


A neuron is adapted to carry an electrical signal through the body and pass it on to adjoining neurons.

from the dendrites of nearby neurons by tiny gaps (synapses). When the electrical impulse reaches the dendrites, it causes them to release a special chemical (neurotransmitter), which crosses the synapse and binds to the cell membranes on the next neuron's dendrites, sparking a new electrical impulse. In this way, an impulse can travel between the brain or a ganglion to the furthest extremities of the body.

Nerves that leave (or arrive at) the brain and ganglia are thicker, formed from bundles of many neurons. In more peripheral body parts they become thinner, containing fewer neurons. Each sensillum may be innervated by just two or three individual neurons.

An ant's brain, though obviously tiny, can make up as much as 15 percent of its total body weight.

# INSECT INTELLIGENCE

**With a brain the size of a sesame seed, a bee seems an unlikely candidate for exceptional intelligence. Yet bees and other insects can perform complex behaviors, suggesting brainpower beyond their size.**

Ⓐ The degree of organization and cooperation required for these ants to form a living bridge for their comrades suggests real intelligence.

Modern research on various animals, particularly birds, has overturned our traditional view that brain size is the only important factor that determines intelligence. A raven and a chimpanzee show comparable intelligence, because the raven's ½-ounce (15g) brain has a much higher density of neurons than the chimp's, and a more efficient structure generally. What about insects, though? Their brains are vastly different from those of vertebrates, and are relatively little-studied.

Defining intelligence is scarcely straightforward. However, most biologists agree that its elements include memory, innovative behavior, ability to learn, planning, problem-solving, and social interaction (including teamwork). All of these can be observed in certain insects, but probably most of all in the social species, such as bees, wasps, and ants.

Honey bees have to learn how to reach the nectaries of various flower types, but they can also be taught skills that clearly are not essential to their survival, such as recognizing different human faces. The famous figure-eight "waggle dance," performed in the colony to tell other bees where to go for new nectar sources, conveys information about direction, distance, and even quality of the nectar available. This use of symbolism to express concepts is a sort of language, the like of which is not known in many apparently "higher" organisms.

Ants' social structure can show great complexity. Some species are "slave-makers," stealing ant pupae from other species' nests to replenish their own workforce. They are

capable of cooperative behavior, such as forming living bridges to allow others to traverse gaps, and working together to carry large prey items back to the nest.

**PREDATORY MINDS**

Carnivores are often smarter than herbivores, because it takes more brainpower to capture an animal than it does to find and eat a plant. Dragonflies have been shown to be capable of predicting the flight path of the insect they are chasing, and are able to ignore all distractions (including other insects, when their target is in a swarm) as they home in, making a successful strike more than 90 percent of the time. Curiosity, another aspect of intelligence, is also apparent when a dragonfly hunts—it will even pause midair to examine you before moving on.

Ⓐ Intelligent behavior is most apparent in social insects like bees, and predators like dragonflies.

Ⓥ To chase down living, fast-moving prey takes a finely adapted nervous system.

# INSECT BUILDERS

*From the exquisitely molded little mud nest of a potter wasp to the towering monolith that houses a colony of termites, the insect world includes many master builders.*

Most of the insects that construct shelters, either for themselves or for their young, are members of the order Hymenoptera (bees, wasps, sawflies, and ants). Wasps of the genus *Eumenes* are among those that craft particularly pleasing structures from mud, which, when complete, resemble dainty vases with a large bowl narrowing to a small entrance passage. The female wasp collects mud, which she sticks to a substrate (often a high vertical surface such as a wall or tree trunk, for protection). She sculpts and manipulates her building material with her mouthparts and forelegs, using her saliva for moisture where required. When it is complete, she provisions it with some paralyzed prey, lays an egg inside, and may then seal up the entrance. The egg hatches and the larva lives on its food store, pupating inside and then breaking out after it emerges from its pupa.

Termite mounds may be tall and narrow, or wide with multiple humps. They include shafts or "chimneys" for ventilation, and a maze of tunnels and chambers inside, leading to the main nest chamber in the ground beneath, where the king and queen termite live and breed. They are built and maintained by worker-caste termites, while the larger soldier caste defend the nest from invaders. As the mounds are built from a considerable amount of termite droppings, they encourage a different and often richer plant community than the surrounding terrain. Clusters of mounds in close proximity effectively create different biomes, and greatly increase biodiversity.

Ⓐ Social insects make their nests from materials such as paper from chewed wood (top, wasp nest) and earth (bottom, termite mound).

Some moth caterpillars live socially in large "tents" made from silk that they produce from facial glands.

Depending on the materials they use, caddisfly larval cases can be quite ornate in appearance.

The little-known tropical insect order Embioptera—the webspinners—live permanently in aggregations within webs spun from glands on their feet. Only the adult males develop wings and disperse from the colonies.

Some insects build miniature shelters just for themselves. The most famous of these are the larvae of caddisflies. These delicate-bodied, slow-moving larvae are aquatic and feed on detritus, so protection is more important than speed. They form their cases from tiny fragments of gravel, twigs, sand, and other materials they find on the lake bed, spun together with silk.

## Webs

It is not unusual in some areas to come across a shrub or hedgerow that is completely draped in what look like spider webs. In fact, these large silk webs are the work of young moth caterpillars, such as the ermine moths (genus *Yponomeuta*). The young larvae spin their web out of silk secreted from their mouthparts, and under its shelter they can feed on the plant's foliage in relative safety. Some other insect larvae use silk to make individual shelters. Moths of the family Tortricidae are sometimes known as leaf rollers, as they roll up and secure a section of a leaf of their food plant and feed within this makeshift shelter.

4

# MOVEMENT

Flight was an insect innovation—insects were the first flyers on Earth and are one of only four animal groups ever to have evolved the ability to self-propel through air. Their other means of locomotion are no less impressive—among their numbers we also find sprinters, leapers, climbers, water-treaders and deep divers.


4.1 • Muscular system

4.2 • Movement on land

4.3 • Flight

4.4 • Swimming and diving

4.5 • Escaping danger

4.6 • Immobility


The jointed legs and articulated bodies of insects allow them great freedom of movement to run, climb, leap, and fight.

# MUSCULAR SYSTEM

**Insects are capable of some truly jaw-dropping physical feats relative to their body size, thanks to an array of efficient muscles with a very high energetic output.**

From the weight-lifting abilities of ants and beetles to the prodigious leaps of fleas and grasshoppers, insects pack a lot of power into their small bodies. Muscles are not only used to drive the legs and wings, mouthparts, and antennae, but they also help push fluids through the body and push food through the digestive tract.

All insect muscle is of the type known as striated, because of its banded appearance visible under the microscope. It works by contraction (shortening). For example, the muscles that attach the top of a leg to the underside of its thorax segment will move that leg as a whole unit when they contract. There are also muscles attaching the body segments to each other, which allow legless insect larvae to wriggle along through a squeezing and stretching action.

An individual muscle is made of several bundles of long, fibrous muscle cells (myocytes). Within the myocyte are alternating clusters of two types of protein filaments—actin and myosin—which are held together by chemical bonds. When the myocyte receives stimulation from a neuron, this causes a chemical change inside the myocyte that breaks the bonds holding the actin and myosin filaments to each other. When the bonds break, the filaments slide over one another, causing the myocyte to shorten in length. Because this occurs across all of the myocytes in the muscle, the entire muscle contracts.

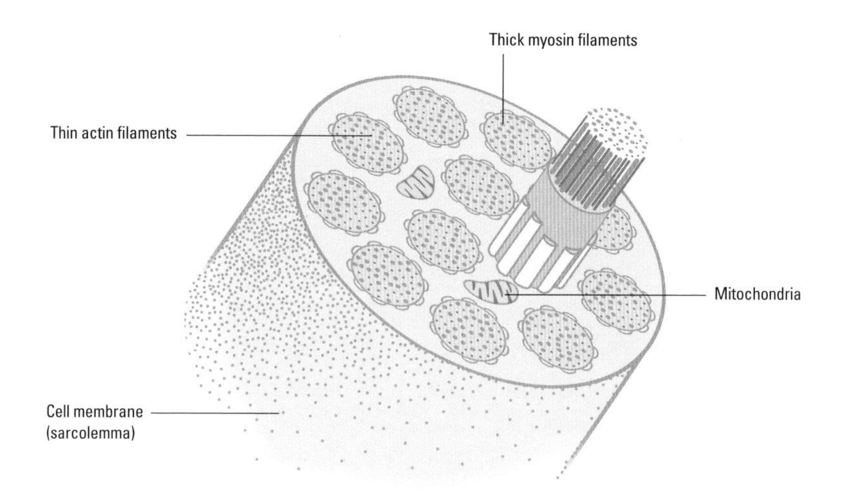


A muscle cell or myocyte shows an organized structure, with the elongated filaments all collected together into bundles and functioning as a single unit.

Ⓐ Worker ants are expert at fetching and carrying, and can lift remarkably heavy loads.

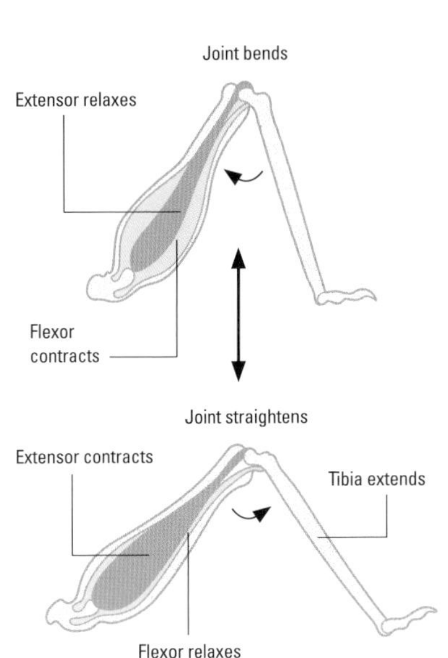


Ⓐ Contraction of flexor muscles causes the joint to bend, while contraction of extensor muscles straightens it.

**FLEXING AND STRAIGHTENING**

Muscles to move a leg joint come in pairs, as is the case with our own muscles. The joints work like hinges and only bend one way. One end of each muscle attaches to each of the two segments either side of the joint. The muscle on the inside of the natural bend is an extensor—when it contracts and shortens it makes the joint straighten. The other muscle is a flexor—its contraction causes the joint to bend.

# MOVEMENT ON LAND

**Many insects—including the beetles, the most successful and diverse of all insect groups—get around on foot. Leg propulsion includes walking, scuttling, climbing, and leaping.**

Insects with prominent strong legs are likely to be walkers or climbers, or they may use their legs to hunt, or to cling to difficult perches. The tiger beetles (subfamily Cicindelinae), which chase down and kill other insects on the ground, are among the fastest runners of all land invertebrates, with *Cicindela hudsoni* of Australia able to race along at 5.6 mph (9kph). The larger cockroaches are also fleet-footed, with *Periplaneta americana* able to reach 3.4 mph (5.5kph). Insects that are runners (cursorial) tend to have long legs with sturdy femurs, and the tarsi are tipped with strong claws for grip. The hind pair are longest but the middle and front legs are also substantial.

The most efficient running gait is "tripod," with three feet in contact with the ground at any moment (the fore- and hind-legs on one side and middle leg on the other). This same gait is used when climbing a vertical surface. However, it has been observed that cockroaches can increase their running speed by taking alternating strides with all three legs on each side.

Grasshoppers and crickets have adaptations to their hind legs to power their leaps. The hind leg is much longer than the others, with an oversize femur to power the strong muscular contraction that fires them forward. They are generally slow and awkward on their feet when moving through vegetation.

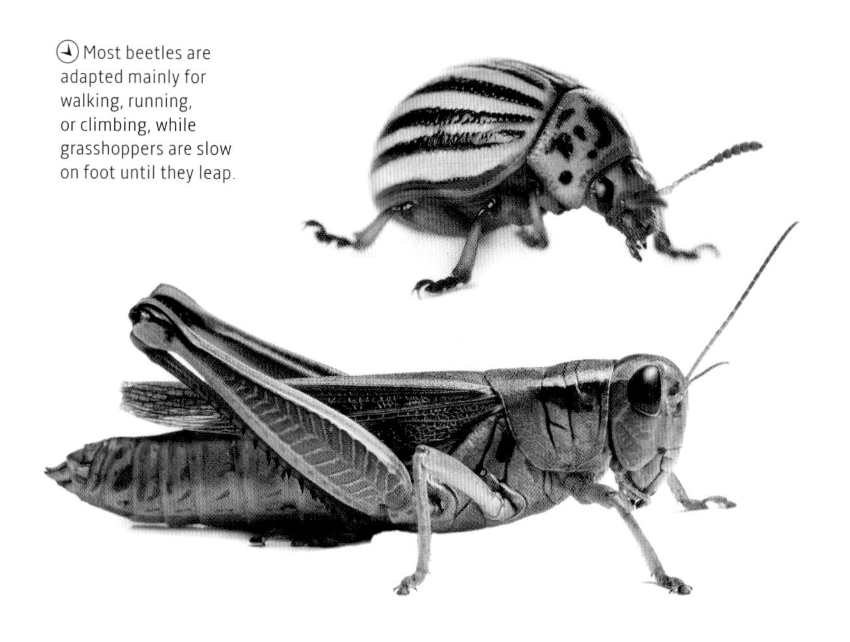

Most beetles are adapted mainly for walking, running, or climbing, while grasshoppers are slow on foot until they leap.

Some grasshoppers can leap more than 200 times their body length.

## STICKING AROUND

A housefly can easily perch and walk upside down on a ceiling, as well as vertically on a glass windowpane. Many other insects are capable of similar gravity-defying feats, thanks to anatomical tricks (coupled with their very low body weight). It's tempting to assume that they have suckers on their feet, but in the case of flies the grip is provided by masses of tiny setae (hairs) on the underside of the tarsus tip, which grip on to equally tiny ridges and fissures on the apparently smooth surface. Some other insects also secrete a sticky substance from their feet to aid the grip provided by the setae.

Flies can perch upside down with ease, even on quite smooth surfaces, thanks to the hairs on their feet.

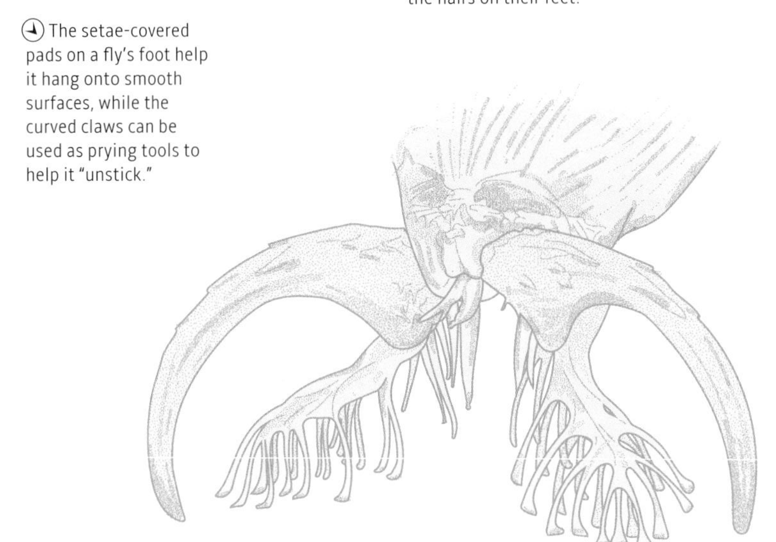

The setae-covered pads on a fly's foot help it hang onto smooth surfaces, while the curved claws can be used as prying tools to help it "unstick."

# FLIGHT

**Insects have been on the wing for 400 million years. Their flight both impresses and baffles us, because its mechanisms and underlying anatomy are so different from those involved in vertebrate flight.**

Direct wing-flapping (dragonflies, damselflies, and mayflies)

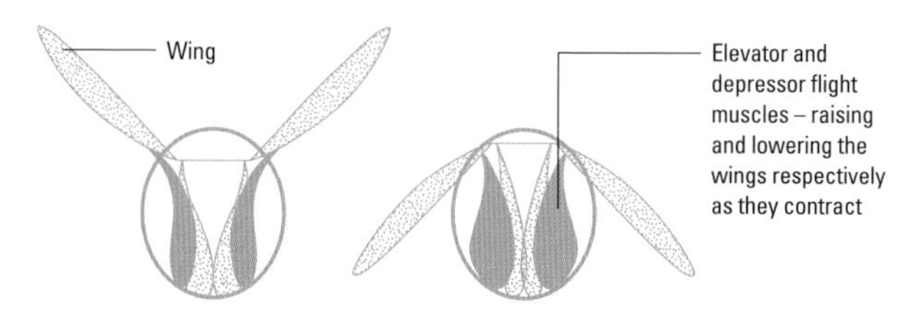


Indirect wing-flapping (other flying insects)

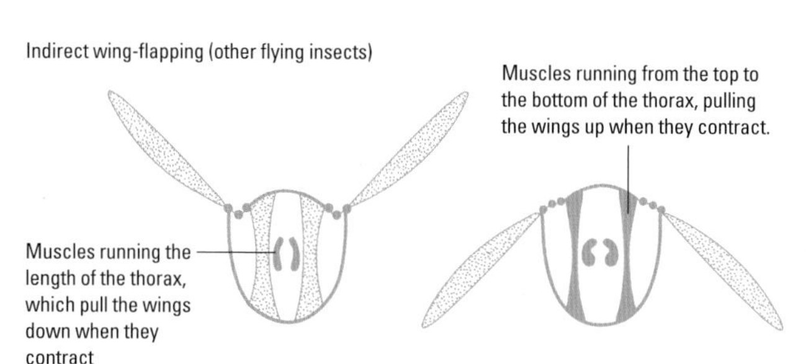


Insect flight muscles are attached to the base of the thorax. The upper attachment point may be the bases of the wings themselves, so the wings move up and down as the muscles relax and contract (top), or to the top of the thorax, so the wings move as a result of the thorax itself flexing and expanding with the muscular contractions (bottom).

Birds flap their wings by using muscles that connect their chest to the "upper arm" bones in their wings. In most insects, though, the flight muscles do not attach directly to the wings at all. Dragonflies and damselflies do have flexor and extensor muscles in their middle and hind thoracic segments that are connected to the bases of the wings. Each pair of wings can be moved independently, which gives these insects great control and maneuverability in the air. Mayflies also have flight muscles stretching between the thorax and the wings.

In most other flying insects, wing-flapping is achieved indirectly, through muscles within the thorax, attached to the upper-side and underside of the segments. As they contract and lengthen, the muscles distort the shape of the entire thorax, which causes the wings to move. This oscillating action is fast enough to drive wingbeats as fast as 62,760 a minute (in the case of some *Diptera* species). There are tiny muscles within the wing as well, which alter its angle of attack.

It is often stated that bumblebees are "scientifically" incapable of flight, because they are too heavy for their small wings to lift them. This (demonstrably wrong) idea assumes that insects fly with up–down wingbeats. In fact, the flight stroke is more circular and involves the wings pushing forward, front-on, then tilting over and pulling back with the full face of the wing

Ⓐ Many species of hawkmoth are expert at hovering, using this ability to feed from flowers that would not support their weight.

against the air. The action also generates tiny vortices (like tornados) in the air above the wing, which reduces air pressure, helping the insect overcome its own weight.

Some insects, including many moths, have a coupling mechanism that holds the front and hind wing on each side together, so that they function as a single pair. This improves energy efficiency, though makes for a less agile flight than in insects with independently moving wings. The way the wings fix together varies—in some cases there is a hook and corresponding notch, in others the wings are not physically attached but the forewing overlaps the hind wing, which holds them together on the flight downstroke.

## CLAP AND FLING

Some very small flying insects, such as thrips, use a very different technique to become airborne. They clap their wings together above their backs, and the downward rebound creates a vortex over each wing that sucks the insect upward. This flight method is known as "clap and fling." Because its action is quite violent and can quickly cause damage to the wings, the technique is most often seen in insects with short adult life spans.

Ⓥ Thrips are among the few insects that fly though a rather violent "clap and fling" action.

# SWIMMING AND DIVING

**Many insects live in water in their larval stages, and some, such as water beetles and water boatmen, remain water-dwellers as adults—even though they are also capable of flight.**

Although insects' ancestors were thought to be marine and freshwater crustaceans, the rather small number of insects on Earth today that are aquatic as adults have evolved from terrestrial ancestors. Aquatic adult insects belong to two major groups, the beetles and the true bugs, whereas most other members of these groups are landbound in all their life stages. Insects that are aquatic only as larvae are more diverse. They include mayflies, stoneflies, dragonflies, caddisflies, and some true flies, lacewings, and moths.

To live in water doesn't necessarily require an insect to swim—it may walk on the substrate, or clamber in underwater vegetation. However, some insects are true swimmers and divers, and possess modified leg anatomy to propel them through water. In most cases, this involves a dense unidirectional "paddle" formed by setae (hairs) on one or more of the leg pairs. In the whirligig beetles (family Gyrinidae), the middle and hind legs are short and flipper-like, while the front legs are long and used for seizing prey rather than swimming. In most cases, water beetles' swimming legs work together rather than alternately.

Among the water boatmen (family Corixidae) and backswimmers (family Notonectidae), both of which are true bugs (Hemiptera), the hind legs are elongated and fringed with hairs. These legs act like oars to move the insect through the water.

Some water insects only swim on the surface, but diving beetles have good enough propulsion to dive deep underwater. They overcome the other challenge of subaquatic

Swimming insects such as backswimmers (left) and diving beetles (right) propel themselves by using their legs as paddles or oars.

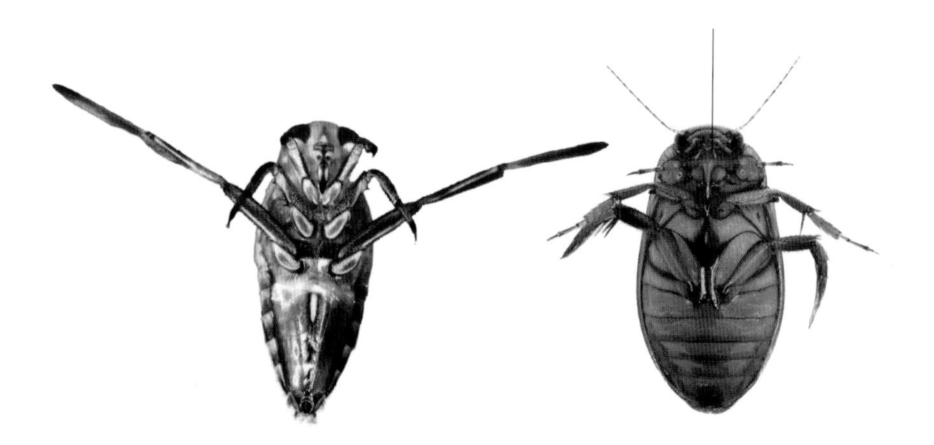

Ⓐ A pond skater's feet dimple but do not break the water's surface tension.

Ⓢ Bubbles caught in the body hairs provide swimming beetles with an underwater air supply.

life—the need for oxygen—by carrying a supply of it with them. They capture a bubble of air under their elytra and breathe from it as they swim. The bubble can also partly replenish itself, as some of the dissolved oxygen in the water will diffuse into it, making it a functional gill. Other swimming beetles and bugs have a layer of breathable air trapped in hairs on their exoskeleton—known as a plastron, this can in some cases supply them with a lifetime of air, meaning they need never surface.

**WALKING ON WATER**

Pond skaters or water striders (family Gerridae), and some other water bugs live on rather than in the water. By distributing their tiny body weight over a wide area, these long-legged insects are supported by the water's surface tension. The tips of their tarsi are covered in waxy hairs that trap tiny air bubbles, providing buoyancy.

# ESCAPING DANGER

**Some kinds of insect movement are only deployed at times of necessity. When danger threatens, it's helpful to have a "special move" that can get you out of trouble in a hurry.**

A huge array of animals prey on insects—including other insects. Even the most formidable dragonfly or hornet will probably end its days prematurely, as lunch for a bigger hunter. But many insects have ways to try to physically escape from danger, beyond simply fleeing as quickly as they can.

The click beetles (family Elateridae) do not have the legs of a specialized jumper, but they can leap nonetheless. They can run and scramble perfectly well, but if they are knocked on their back they can only right themselves by "clicking"—and the same move can help them escape a predator. The front segment of the thorax has a small projection at its rear. When the insect flexes its body swiftly and strongly enough, this projection is locked into a corresponding notch on the second thoracic segment. The recoil force propels the beetle into the air with an audible click. It may turn several somersaults before landing some distance away and (with luck) the right way up.

Some moths and butterflies use "startle coloration" to try to dissuade predators from attacking them. The Eyed Hawkmoth (*Smerinthus ocellata*) has colorful eye-like markings on its hind wings. It rests with the forewings covering the hind wings. If it is attacked while at rest, it uses its wing muscles to move the front wings forward rapidly. This flashes the hind wing "eyes" and hopefully distracts the predator.

Many predators will lose interest if their prey appears to suddenly drop dead.

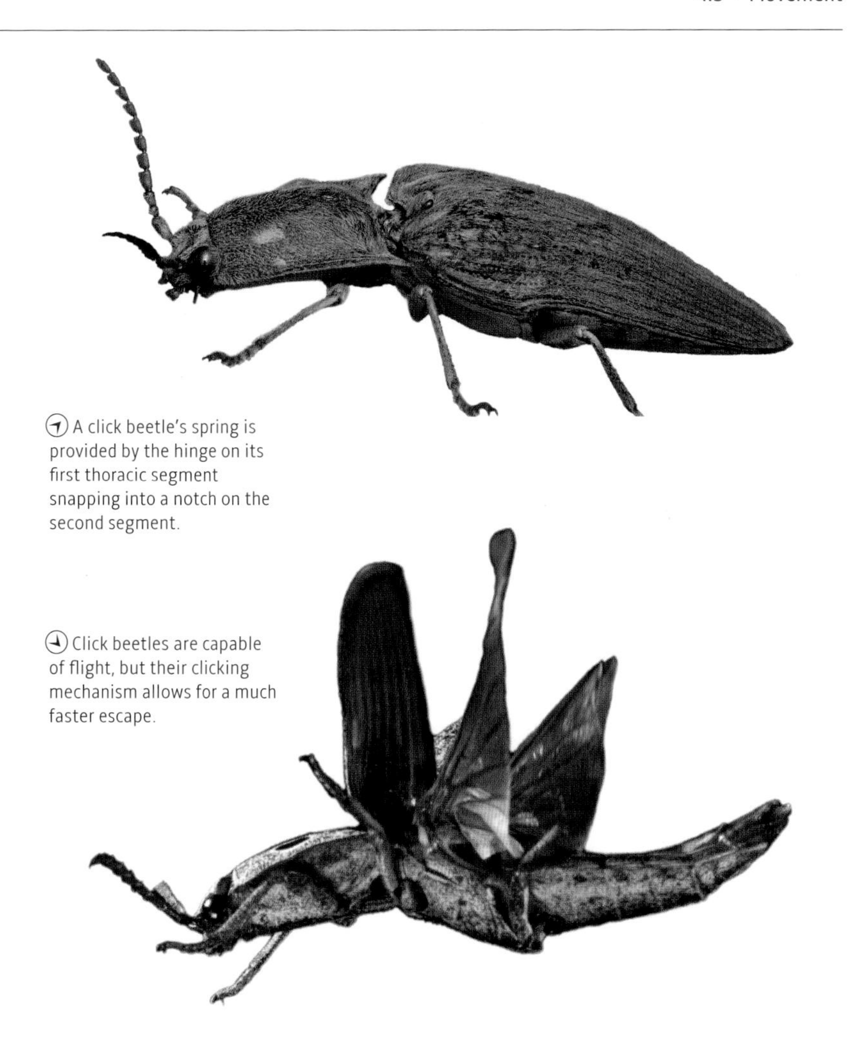

A click beetle's spring is provided by the hinge on its first thoracic segment snapping into a notch on the second segment.

Click beetles are capable of flight, but their clicking mechanism allows for a much faster escape.

**PLAYING DEAD**

Many foliage-dwelling insects, if disturbed, will release their grip and fall limply to the ground, hoping that the leaves will hide them wherever they land. If they have fallen as far as they can and the potential predator is still interested, they may continue to "play dead" for a long spell. This is effective self-defense, as predators only attack live prey. Some ground-dwelling insects also play dead very effectively—certain beetles of the family Tenebrionidae (the darkling beetle) are known as "death-feigning beetles," so readily and effectively do they take on a lifeless appearance.

# IMMOBILITY

**The class Insecta includes some of the world's most well-traveled migratory animals, but there are also some insects that, from hatching to death, barely move at all.**

Animals that do not move are known as sessile. Many of them begin life as highly mobile—for example, the acorn barnacles that encrust shoreline rocks start out as free-swimming larvae, before they find a place to settle in the intertidal zone and secrete the calcified plates that surround them and hold them in place for the rest of their lives.

There are few truly sessile insects. Scale insects (superfamily Coccoidea) do qualify, or at least the females do. These usually very small members of the true bug group feed by sucking the sap out of plant stems with their piercing mouthparts. They are mobile on hatching, but after their first few molts the wingless females become fixed to one spot, feeding continuously, and typically secrete a thick shield of wax for protection. Males usually do have wings on maturity, enabling them to find and mate with females. Some scale insects are guarded by ants, which gather the honeydew they excrete. The ants may even move young scale insects around their host plant, to suitable feeding spots.

The females of several species of moths are flightless and almost immobile. The female Vapourer Moth (*Orgyia antiqua*) has a very distended abdomen and reduced, stumpy wings. When she emerges from her pupa, she releases pheromones that attract males, and after mating she lays her large clutch of eggs on and around the silk cocoon that had protected her while she was pupating. Vapourer caterpillars are very

Ⓥ Ladybugs' bright coloration warns of their distasteful bodies and discourages predators from attacking them.

mobile and active, and make the most of this ability to find a safe site to pupate where they will be able to complete their life cycle away from predators.

Ⓐ A female Vapourer Moth is far more active as a caterpillar than it will be once it reaches its adult phase.

## MOTIONLESS SURVIVAL

Even the most active insects may spend hours or days motionless, under certain conditions. In temperate parts of the world, many adult and larval insects pass the coldest months in a state of hibernation. With drastically slowed metabolism, they conserve their energy so that they can resume activity when the temperature rises. Even in mid-summer there may be sustained days of rain or low temperatures, forcing them to become inactive. This is why even highly active insects generally have good camouflage, or if not they have highly unpalatable bodies and exhibit bright coloration to warn of this—enforced immobility is a part of life for them.

Ⓐ Mealybugs and other scale insects feed by sucking fluids from plant tissue, and many species barely move throughout their lives.

5

# FEEDING AND DIGESTION

Most insects are herbivores, but a sizable minority are scavengers or hunters. Some are adapted to suck up liquid food, others to bite and chew their way through solids, and their mouthparts and digestive tracts are adapted to handle everything from nectar and blood to flesh and solid wood.

5.1 • Mouthpart anatomy

5.2 • Types of diet

5.3 • The digestive tract

5.4 • Processing food

5.5 • Changes during life cycle

5.6 • Drinking and fluid balance

Caterpillars are eating machines, consuming thousands of times their own body weight during their few weeks of activity between hatching and pupation.

# MOUTHPART ANATOMY

**Many of the major insect groups can be distinguished by (among other things) the structure and function of their external feeding equipment—their mouth appendages or mouthparts.**

The human diet is diverse, and in our mouths we have all the tools we need in one place to cut up, grind up, lick up, or suck up whatever we want to consume. Most insects, though, specialize in one style of eating only, and have mouthparts adapted for that purpose.

The mouthparts originate as appendages from the six head segments and sclerotized plates of the cuticle. They include jointed, highly maneuverable parts, more rigid parts, and sensory structures (palps). In general terms, insect mouthparts have up to five principal components:

- **Labrum:** a sclerotized plate that functions as an upper lip
- **Mandibles:** paired strong, biting, or crushing appendages that work as jaws
- **Maxillae:** paired, delicate appendages that manipulate food and are attachment points for sensory palps
- **Hypopharynx:** usually a small tongue-like structure that adds saliva to food and assists with swallowing
- **Labia:** paired and fused appendages that function as a lower lip, and also have sensory palps

Insects with this basic arrangement include the grasshoppers and crickets, cockroaches, and some beetles. However, in many more recently evolved insects the mouthparts are highly modified and some elements have been discarded. In the true bugs (Hemiptera), the mandibles and

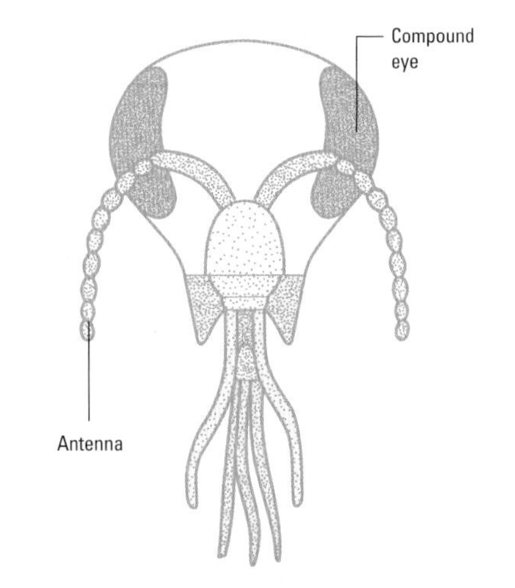


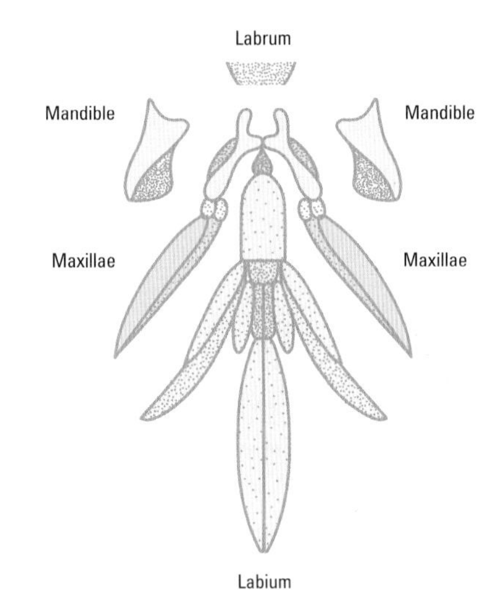


Ⓐ The head of a typical bee, showing the biting and licking mouthparts: The hypopharynx (not shown in this view) contains glands, which in honeybees secrete royal jelly.

(>) The coiled sucking proboscis of butterflies and moths is a flexible tube, formed by the maxillae.

(v) Ants have large, strong mandibles, suitable for biting and grabbing.

maxillae are fused and reshaped to form a firm, sharp-tipped tube with which the insect can pierce plant or animal tissue and suck liquid from it. Adult butterflies and moths lack mandibles, and part of their maxillae are fused and formed into a long, flexible sucking tube (proboscis), which when not in use is coiled up between the enlarged sensory palps of the maxillae.

**LARVAL MOUTHPARTS**

Because an insect larva does little but feed and grow, its mouthparts are often more obvious and much more heavily used than those of the adult. The majority have fully functional mandibles and eat by biting and grinding their food. The larvae of dragonflies and damselflies are notable in having a spine-tipped labium that can "fire" forward at high speed to pierce prey.

# TYPES OF DIET

**Organic matter of all kinds is fair game for at least some kinds of insects. Some are generalists in their diet, with generalized mouthparts to match, while others are extreme specialists.**

At least two-thirds of all insect species are plant-eaters (herbivores). Of the rest, most are meat-eaters, feeding on the bodies of other animals that they kill, parasitize, or scavenge (carnivores), and a relatively small proportion (under 10 percent) eat plants and meat, and are classed as omnivores. In some cases, the larval and adult diets are quite different.

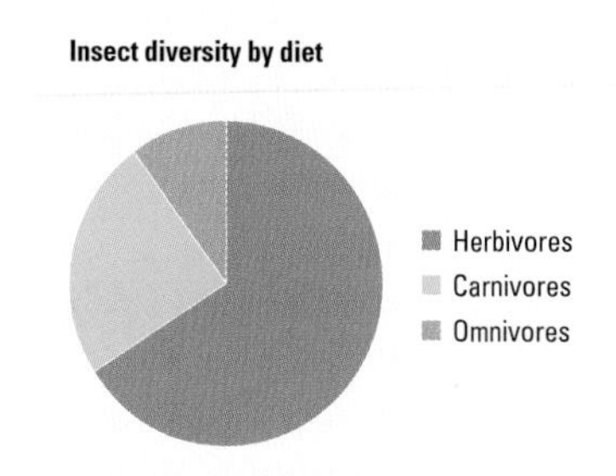


Insects that feed on plants may consume the foliage (typically using mandibles to bite pieces out), suck the sap from stems (as many true bugs do, with their sharp, piercing mouthparts), or take nectar from flowers. Nectivorous species obtain nectar in a variety of ways—for example, using an extensible sucking proboscis, as described on the previous page, in the case of butterflies and moths, or a sheathed, licking tongue formed from parts of the maxillae and labium in bees (which also have mandibles to eat pollen). Some foliage-eaters have a highly specialized diet, only feeding on a single plant species and only using

Hoverflies and many other true flies have broad sponge-like tips to their labium, which draw in liquids.

those individual plants that are growing in specific conditions.

Insects that are carnivores may capture live prey, as dragonflies and mantises do, eating prey by taking out bites with their mandibles. Most predatory insects catch other insects, though there are documented cases of large mantises catching tiny vertebrate prey, such as hummingbirds. Omnivorous insects include some members of the order Orthoptera, such as the Great Green Bush Cricket, *Tettigonia viridissima*, which preys on other insects but also chews up plant foliage. Other omnivores feed on dead or decaying plants, and/or animals.

Ⓐ Weevils are herbivores, and may be hunted by predators like ants.

## BLOODSUCKERS

Drinking the blood of vertebrates is a lifestyle followed by a small number of insects. Fleas are flightless parasites that usually have a particular host species (though may at times feed from another, as many cat and dog owners have found out to their cost). There are also many blood-drinkers among the true flies, such as mosquitoes and horseflies, the females of which feed on blood from any suitable host they find. They may also eat other foods such as nectar, but they are unable to form their eggs without the protein they obtain by consuming blood. Blood-drinking insects have anticoagulants in their saliva, so the blood from the wound they make does not clot before they have finished feeding.

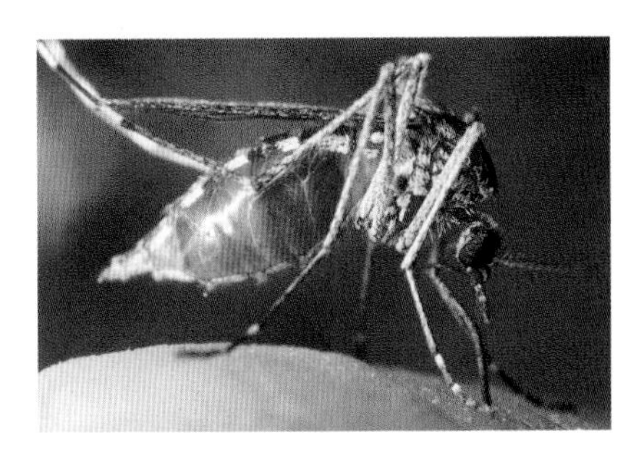

Ⓐ Mosquitoes and some other true flies feed on the blood of vertebrate animals.

Ⓐ Caterpillars consume large amounts of plant foliage.

# THE DIGESTIVE TRACT

**The food that insects and other animals eat needs to be processed to have valuable nutrients extracted, before what is left is excreted. This is the job of the digestive tract.**

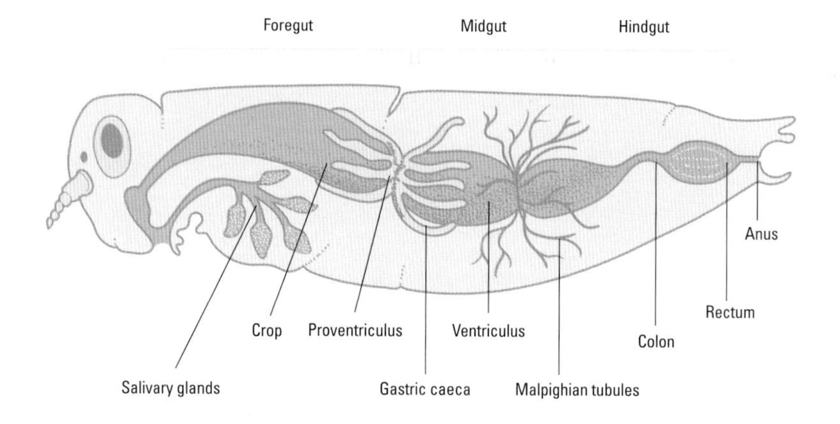


Ⓐ The insect digestive tract has three main regions. Food breakdown occurs mainly in the foregut, food absorption through the mid- and early hindgut, and reabsoprtion of water in the hindgut.

Some invertebrates, such as flatworms, take in food and excrete waste through the same opening, but the insect digestive tract, or alimentary canal, is a so-called "complete" digestive system, with food taking a one-way journey from mouth to anus via three main functionally distinct regions, the foregut, midgut, and hindgut.

Food is mixed with saliva in the insect's mouth, to aid swallowing and begin digestion. Each pair of mouthparts has an associated pair of glands (though not all are present in all insects), some of which produce saliva. This is carried to the mouth via ducts. The food then passes into the first part of the alimentary canal, the foregut, through the action of the cibarial muscles around the base of the mouth, which contract to create suction that pulls the food in.

The foregut contains the crop, an expandable part of the tract where food is stored prior to further processing. This organ allows the insect to consume a meal quickly, and some digestion via the salivary enzymes may occur. Digestion begins in earnest when the food passes into the proventriculus. This is a muscular organ, lined with sharp projections that mechanically break up food.

In the midgut, food reaches the ventriculus, where it is doused in digestive enzymes, and nutrients are absorbed through tiny projections (microvilli) on the cells lining the wall. Waste removal takes place in the hindgut, where numerous very fine tubes (Malpighian tubules) branch off the gut. They remove ammonia from the insect's

All insects that consume food will excrete some form of waste material.

hemolymph (its equivalent of blood) and return it in the form of uric acid to the hindgut for excretion. The other main function of the hindgut is to reabsorb water and salts from what remains in the gut, before the waste is excreted via the rectum.

**INNER SKIN**

The inside of the crop and some other parts of the gut are lined with a very thin layer of cuticle—the same material that forms an insect's exoskeleton—known as intima. Just as it works on the outside of the body to protect it from damage, the intima helps protect internal parts from being damaged by hard food particles.

Leaf-mining insects, in this case the larvae of the moth *Stigmella aceris*, eat their way through the cells within a leaf, the "mine" they create growing wider as they grow. Their droppings, or "frass," can be seen within the mine as a dark trail.

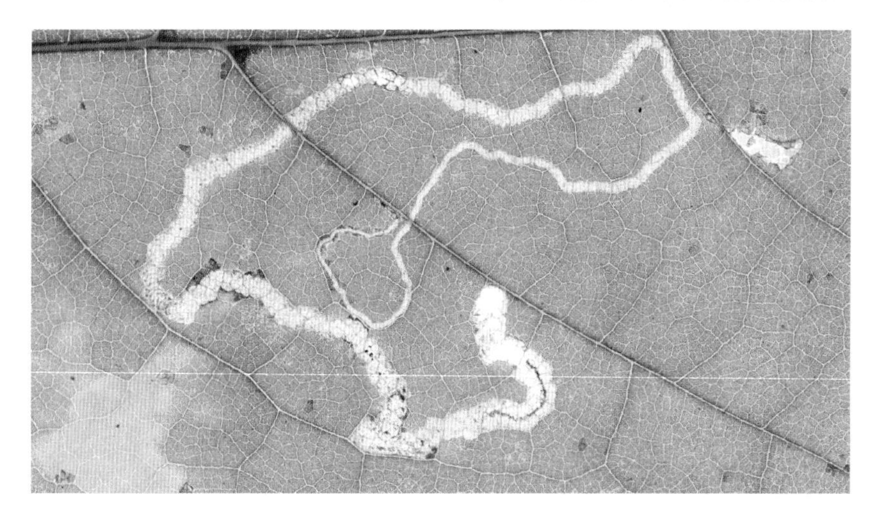

# PROCESSING FOOD

**To survive, insects need energy, and to grow and develop, they need building materials. Both are obtained by the breakdown and rebuilding of the organic material that they eat.**

The main constituents of the food we (and other animals) eat are proteins, fats, and carbohydrates. All are needed to build body tissues, and all (but especially carbohydrates) are potential sources of energy too. However, when initially consumed they are usually in the form of large, complex molecules. Digestion involves breaking them down into very small, simple molecules, which may then be recombined into different larger molecules as required.

Food processing involves mechanical and chemical breakdown. This begins in the mouth—many insects use their mandibles to bite off small pieces of food, and salivary enzymes secreted by the labial glands and hypopharyngeal glands begin to break the chemical bonds holding large molecules together. The enzymes present in the saliva vary according to species—predators' saliva will contain proteases, for breaking down proteins. That of herbivores contains amylases, for breaking down cellulose and other carbohydrates, and may also contain enzymes that help deactivate toxins in plant tissues.

Further mechanical breakdown occurs in the proventriculus, while the salivary enzymes continue to do their work. In the midgut, more enzymes are secreted to continue chemical breakdown. This section of the gut has no lining of cuticle, just a layer of mucous, so small food molecules can pass through and be absorbed by the microvilli of

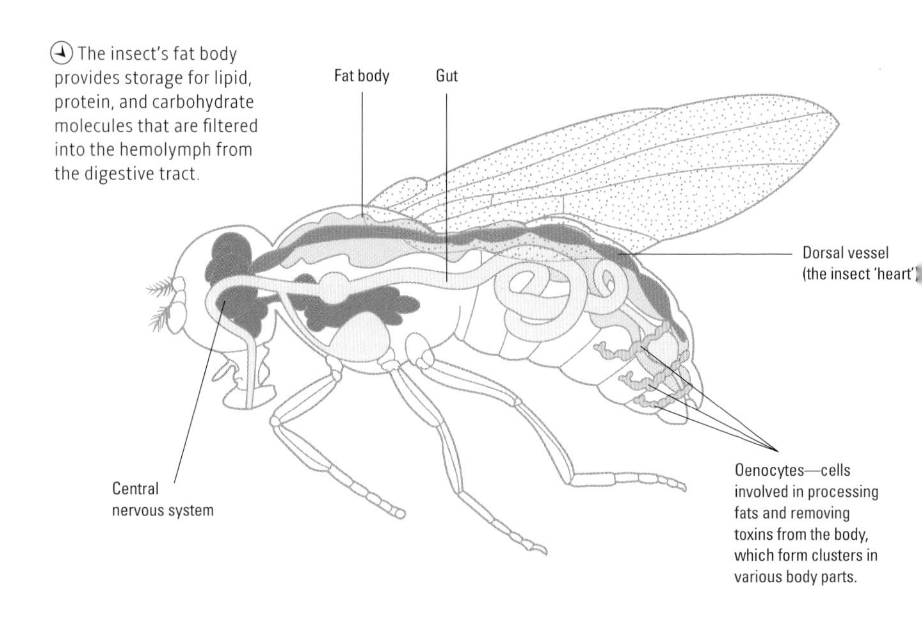


The insect's fat body provides storage for lipid, protein, and carbohydrate molecules that are filtered into the hemolymph from the digestive tract.

Ⓐ The Goldenrod Gall Fly larva's secret to its survival through freezing winters may lie in its fat body.

the cells lining the wall of the gut. Once within the cells, they may be used there and then, or pass via the hemolymph system to other parts of the body. Digestion of nutrients is almost complete after this, but absorption of some salts takes place in the hindgut.

## MINI MOLECULES

The long, chain-like molecules of fats, proteins, and carbohydrates can all be broken down to much smaller and simpler molecules, the "links" of the chains. In the case of fats, these are fatty acids. With proteins, the constituent building blocks are amino acids, and with carbohydrates they are simple sugars—especially glucose. From these materials, insects can synthesize new cells for every kind of body tissue, as well as enzymes and hormones. Glucose and fats provide a long-term energy store in the form of larger molecules—glycogen and triglycerides. These are synthesized and stored in adipocyte cells within a specialized tissue called the fat body. This works in a similar way to the liver in vertebrates, with multiple functions including building and storing energy reserves.

The fat body may have another special role in certain insects. The Goldenrod Gall Fly, (*Eurosta solidaginis*), a North American species, spends the winter in its larval form, inside a sheltered crevice. The shelter is not enough to keep it from freezing almost solid, which would usually fatally damage the cells in its fat body. However, the fat body of this larva contains an unusual type of fat molecule (acetylated triacylglycerols) which is believed to remain liquid at sub-zero temperatures, protecting the cells from damage.

# CHANGES DURING LIFE CYCLE

**Once a larva is fully grown, it undergoes metamorphosis—it transforms to its adult form, in some cases via a pupa stage (see chapter 9). This often entails a drastic change of diet, methods, and feeding anatomy.**

For many insects that undergo incomplete metamorphosis, the transition from larva to adult involves a less radical bodily transformation and the diet and feeding apparatus changes less drastically than in those insects that pass through a pupal stage. Within the order Orthoptera (grasshoppers, locusts, and crickets), for example, diet and feeding methods usually change very little through development. In the case of Odonata (dragonflies and damselflies), both larvae and adults are predators, but their hunting methods are quite different and their mouthparts are somewhat different too. The aquatic larvae stalk prey on foot and spear it with their remarkable extensible labium, which fires out at high speed. The winged adults lose this adaptation to the labium. Instead, they capture prey in flight, "netting" it with their bristly legs. If the victim is too large to be eaten on the wing, they keep hold of it with the forelegs and land on a perch with their four free legs.

Among insects that have a complete metamorphosis, the feeding anatomy usually undergoes considerable reshaping during pupation, changing from biting and chewing mouthparts to a more specialized form. Larvae require more food, and food with a higher protein content, than adult insects. Adult insects of some groups feed only on nectar, which provides sugars for energy and almost nothing else. Some do not feed at all but survive on fat stores accumulated in their larval stages. However, adult insects may require certain foods that as larvae they did not need—for example, adult female mosquitoes need to consume some blood to form their eggs, and adult male butterflies of some species need particular salts to produce sperm. Bees of the family Halictidae are known as "sweat bees" as they drink the salty perspiration and tears secreted by mammals such as ourselves.

Ⓐ Adult dragonflies are superbly adapted aerial hunting machines—very different in appearance from their aquatic larvae.

Ⓐ Dragonfly nymphs are slow stalkers rather than high-speed chasers but are just as predatory as the adults.

Cinnabar Moth caterpillars sequester toxins in their bodies from the plants they eat. These toxins are retained in the adult form.

The adult Cinnabar retains these toxins in its body, and, like its younger self, exhibits bright warning coloration.

**TOXIC STOCKPILE**

Herbivorous insect larvae have ways of dealing with the toxins that some food plants produce. They might break down the toxins, or alternatively they may sequester or store the toxins intact in their own body tissues. This makes the larva's own body toxic to any predators that might consume them, and through association predators will learn to avoid similar larvae. For this reason, larvae that sequester toxins usually have a bold and distinctive appearance. The sequestered toxins remain active in the body when the insect transforms into an adult and retain their protective effect, even though the adult insect does not consume anything toxic.

Many moths of the widespread family Erebidae (the tiger moths) sequester toxic alkaloid chemicals in their food, and in some species these same chemicals are then used in adult males to form courtship pheromones.

Adult butterflies consume far less food than caterpillars, but male butterflies need to consume certain minerals in order to form sperm.

# DRINKING AND FLUID BALANCE

**The right fluid balance is essential for survival, and regulating this can be a particularly delicate process for insects with a small body volume, as they are highly vulnerable to dehydration.**

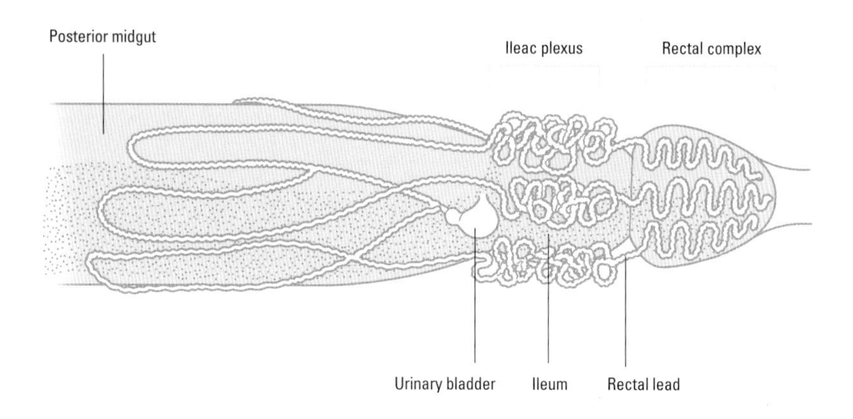


Ⓐ The midgut and hindgut of a caterpillar, showing the Malpighian tubules that extend outward from the gut itself: In the hindgut, the tubules' coils are gathered into two distinct regions—the ileac plexus, surrounding the ileum, and the rectal complex, surrounding the rectum.

For the earliest insects, transitioning to life on dry land could only take place through evolving an excellent system of fluid regulation. A small body can only hold a small volume of water, which will be lost quickly in conditions that encourage evaporation. On the other hand, too much fluid could damage their bodies, as their cuticles are mostly rigid and, in most cases, cannot expand very far. For larvae, taking on extra fluid before each molt expands the body and helps to break the old cuticle, allowing it to be shed.

Another adaptation is that insects are generally able to cope with much wider variations in their internal fluid level than most other animals. This can be tested by looking at the osmotic pressure (see the chart, opposite), a measure of how dilute or concentrated a fluid is, of their hemolymph. In a housefly, this pressure may vary from 4 to 20 atmospheres, depending on how recently the fly had access to drinking water. By contrast, human blood plasma varies very little (from 7.6 to 7.8 atmospheres).

The tough cuticle with its waxy covering helps prevent water loss through evaporation in insects that live in the open. Water loss occurs primarily through the spiracles (see page 94)—these can be closed to prevent drowning in water, but they can also be closed when the insect is in the air, to prevent water loss during strenuous activity such as flight (at such times, the insect's oxygen needs can be met by air held in the air sacs within its tracheal system).

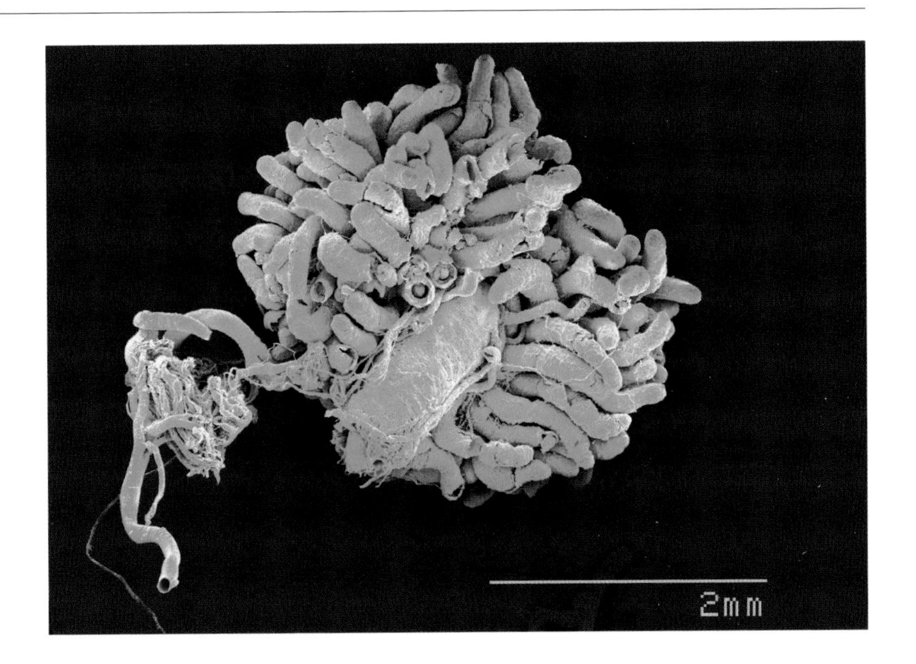


Ⓐ This scanning electron microscope image shows the mass of Malpighian tubules extending from a cockroach's gut.

## REABSORPTION

The Malpighian tubules and other parts of the insect hindgut play a key role in releasing or holding back fluid in the body, with a function analogous to the kidneys in vertebrates. A water-like fluid flows through the tubules. Its low concentration encourages waste products (especially uric acid) to diffuse through from the hemolymph. The fluid passes back into the hindgut, where it is added to the waste products of digestion. Then excess water is reabsorbed in the last parts of the tubules and in the rectum.

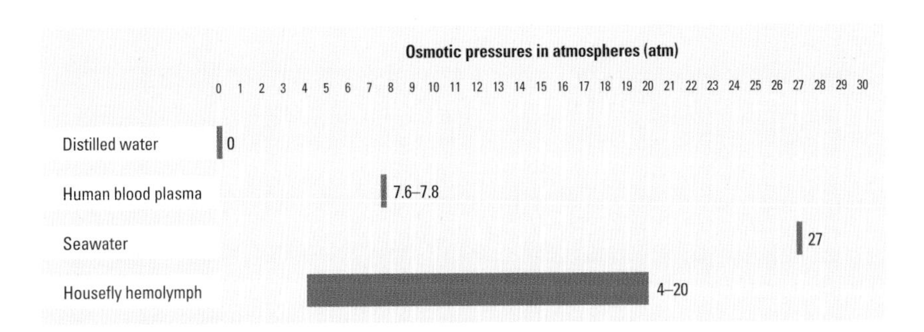

6

# THE RESPIRATORY AND CIRCULATORY SYSTEMS

The elegant way that insects obtain oxygen from the air and pump fluids around their bodies has had a profound influence on the direction of their evolution. In some ways this has been a limiting factor, most notably on body size, but it also creates opportunities not available to larger air-breathing animals.


6.1 • Breathing system

6.2 • Gas exchange

6.3 • Circulatory system

6.4 • Hemolymph

6.5 • Unusual adaptations

6.6 • Hormones


Diving beetles and their fierce, huge-jawed larvae have a range of respiratory and other bodily adaptations to suit their life underwater.

# BREATHING SYSTEM

**Insects have an enormously different breathing system to us—one that offers various advantages but also imposes certain curious limitations that we do not experience.**

Like other animals, insects need to take in oxygen to drive energy-freeing chemical reactions in their cells. They also need to lose the carbon dioxide that is produced by the same process. But while we can only take in and get rid of these gases through our mouths and noses, insects have multiple air intakes over a wide area of their bodies.

In most insects, the middle and last thoracic segments and the first eight or nine abdominal segments each have a pair of spiracles—round openings that sit one on either side of the segment. Spiracles are lined with hairs to trap dust and keep in water vapor to prevent dehydration. They can be fully closed with a valve, keeping the insect from drowning if it falls into water.

The spiracles lead into a system of tubes (tracheae) that branch into smaller passageways, extending through the whole body, delivering air directly to the tissues (rather than via the blood, as in vertebrates). The tracheae are supported and held rigid by rings of chitin, the same material that forms the insect's cuticle. In some insects (particularly the more active, flying species), the tracheal system also includes air sacs, to hold extra air.

Air enters the spiracles passively, though larger-bodied insects can help draw air in through contractions of the abdominal muscles. When the muscles relax after contraction, air is pulled in. This movement can be seen if you look closely—the abdomen of a honey bee, for example, will seem to flatten and puff up in a rhythmic motion.

Spiracles on the body segments of a beetle.

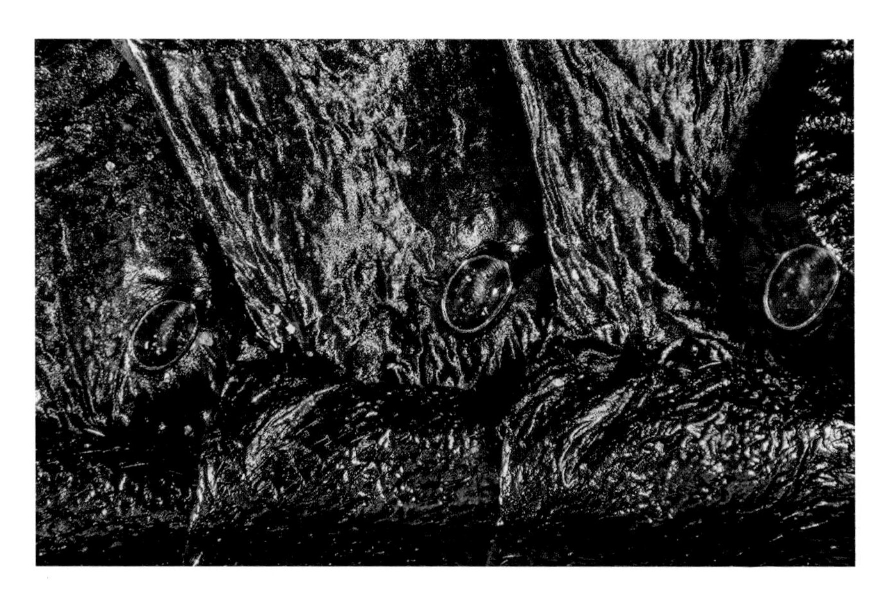

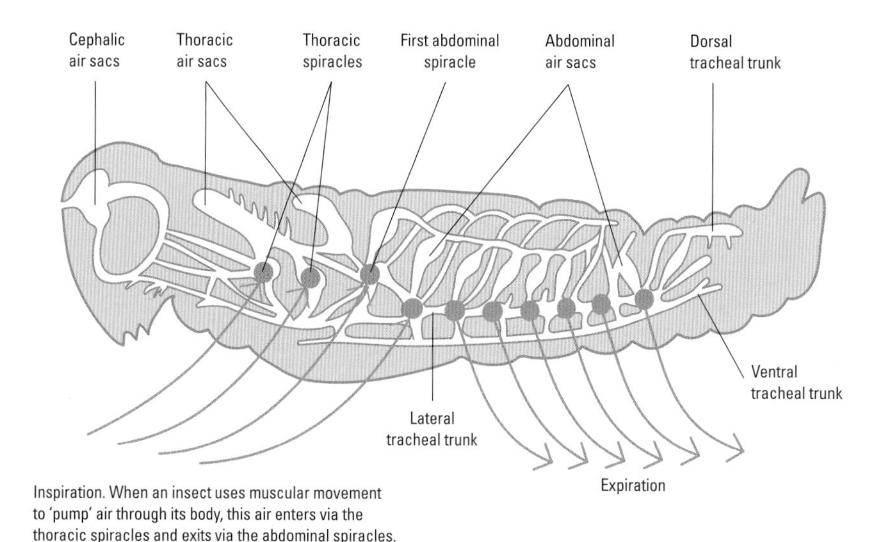


Inspiration. When an insect uses muscular movement to 'pump' air through its body, this air enters via the thoracic spiracles and exits via the abdominal spiracles.

## KEEPING IT SMALL

The way insects take on air is one of the reasons that they must have a small body size. Even with their large number of spiracles, and with assistance from the abdominal muscles, they cannot take in air very quickly. A larger body, with a higher volume to surface area ratio, would need much larger spiracles and bigger muscles to pump the body. This would have knock-on effects—less space available for other body parts, more time needed to molt, much thicker tracheae needed to ensure they could remain open just through air pressure without being squashed by the weight of the body, and so on.

In prehistoric Earth in the Carboniferous Period, when the atmosphere was much more oxygen-rich, much larger insects did exist. However, body size is also limited by the insects' lack of bones—an exoskeleton strong enough to support a much higher body weight would be impractically thick and render its wearer too slow and awkward to survive.

Ⓐ The air system of some insects includes air sacs—spaces where a supply of air can be held, should the spiracles (in blue) need to be closed.

Ⓐ A Giant Weta, heaviest of modern insects. Insect body size is limited by the amount of oxygen in the atmosphere.

# GAS EXCHANGE

**The Earth's atmosphere is 78 percent inert nitrogen, but also contains two gases vital to life—oxygen and carbon dioxide. Animals need oxygen for survival, and plants utilize the carbon dioxide that animals produce.**

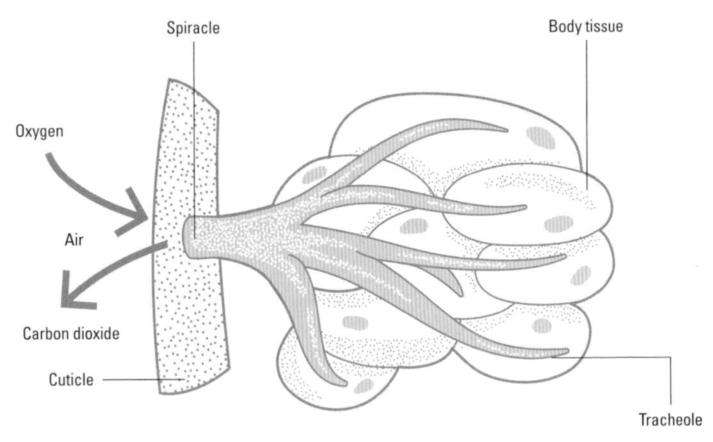


Air entering the spiracles in the cuticle passes along branching tracheoles, which reach the cells forming internal body tissues, and allow direct gas exchange.

Insects, as we have seen, take in air from the atmosphere via the paired spiracles on each of their abdominal segments. Inside their bodies, the air passes through a narrowing and branching system of tubes—the tracheae. The smallest, terminal branches, the tracheoles, are tiny enough to deliver air directly to the cells.

The ends of tracheoles, where they contact the body cells, are fluid-filled, and gases dissolve into this fluid in both directions, with oxygen passing from the air in the tracheole into the fluid, and carbon dioxide passing from the body cells into the fluid. The relative concentrations of the two gases dictate the direction of movement—each diffuses from an area of high concentration to an area of low concentration. Because oxygen is taken in via the tracheae and consumed by the body cells, its concentration is lower in the cells so it diffuses in that direction. Carbon dioxide is produced by the cells, so it diffuses outward, leaving the body via the network of tracheae. The cells also produce water vapor as they consume oxygen, and some of this will also leave the body in the same way, though the insect can recapture it if needed by closing the spiracles.

### THE RESPIRATION REACTION

As with other animals, insects primarily derive their energy from glucose, which comes directly from their diet or from their body's stores of glycogen molecules (which are, effectively, long chains of glucose molecules). Oxygen is required for the chemical reaction that breaks down glucose molecules to release energy.

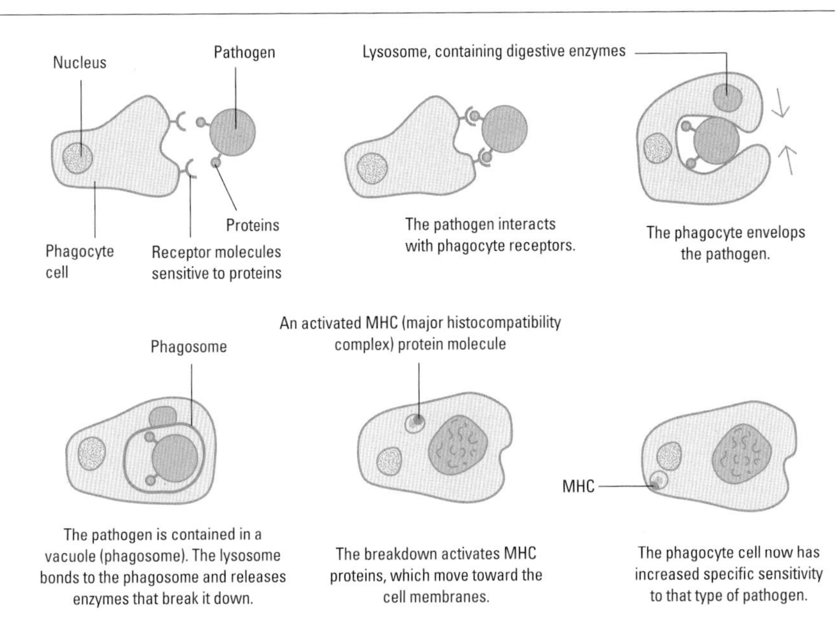


stonefly *Perla marginata*—the hemolymph contains molecules called hemocyanins, which (like hemoglobin in vertebrates' red blood cells) can bind to and transport oxygen. Hemocyanins are common in some other arthropod groups, but in insects the tracheal breathing system has largely replaced them.

When a larva is ready to molt, it may retain extra water in its hemolymph to help fracture its cuticle. After molt, before the new cuticle hardens (becomes sclerotized), the insect may pump hemolymph into certain body-sections to allow them to grow bigger. The final molt from larva (or pupa) to adult involves a period of extensive redistribution of hemolymph, to prepare the insect's body for this new and often very different life stage.

Ⓐ The process of phagocytosis: A phagocyte cell engulfs a pathogen and destroys it, and in the process gains increased sensitivity to that type of pathogen.

## HEMOCYTES

Hemocytes are cells that are carried in the hemolymph. The majority of them are phagocytic, meaning that they can engulf and break down unwanted cells in the hemolymph, such as bacteria and other pathogens, but also dead cells from the insect's own body. Some larval insects have hemocytes that are adapted to attack and destroy the eggs laid by parasitic wasps. See page 186 for a detailed look at the different types of hemocytes and their function in the insect immune system.

# UNUSUAL ADAPTATIONS

**The great diversity of insects, in both body type and lifestyle, has naturally given rise to various specializations in terms of how their circulatory and respiratory systems function.**

In insect larvae of certain species, especially those that live in aquatic or very humid environments, hemolymph can make up nearly 50 percent of the body-weight. In these soft-bodied larvae, the cuticle is thin to allow the body to be highly pliant. However, this means the exoskeleton is not very strong, so instead the body's high fluid content provides much more bodily support. The prolegs on a caterpillar, several pairs of which sit behind the six true legs, have very limited musculature and are primarily powered by the flow of hemolymph.

Ⓐ A caterpillar's prolegs, on the rear half of its body, are held rigid with the aid of hemolymph flow, enabling them to grip strongly.

Several insects, including certain moths, dragonflies, and true flies, have a sophisticated hemolymph flow system to help them regulate their body temperature. They have a muscular sheet on the underside of the body (the ventral diaphragm), which can be used to pump hemolymph from the thorax (where much of the body heat is generated, by the action of the flight muscles) into the abdomen to help the body lose heat.

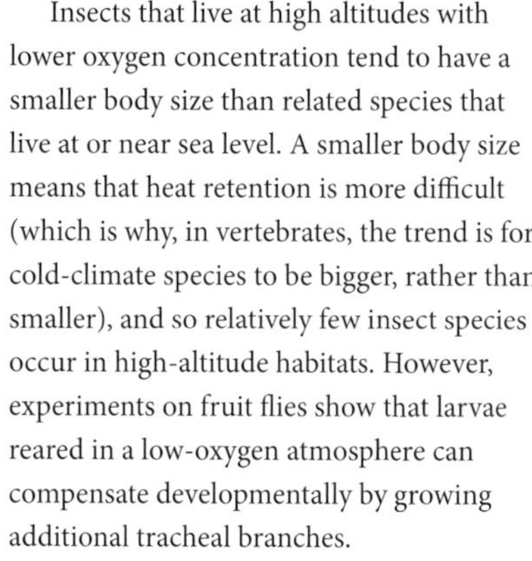

Insects that live at high altitudes with lower oxygen concentration tend to have a smaller body size than related species that live at or near sea level. A smaller body size means that heat retention is more difficult (which is why, in vertebrates, the trend is for cold-climate species to be bigger, rather than smaller), and so relatively few insect species occur in high-altitude habitats. However, experiments on fruit flies show that larvae reared in a low-oxygen atmosphere can compensate developmentally by growing additional tracheal branches.

## BREATHING UNDERWATER

There are several ways that underwater insects obtain their oxygen. Some, such as the "rat-tailed maggot" larvae of certain hoverflies, have a long, narrow tube or siphon at their rear ends, which projects out of the

water into the air. Diving beetles may trap air under their elytra. The larvae of mayflies have feathery gill-like structures through which oxygen can diffuse directly out of the water into the body tissues, while dragonfly larvae have internal gills that absorb oxygen sucked into the body via the rectum. Absorbing oxygen from water is only possible if the water is moving, so mayfly larvae keep a constant flow of water over their gills by fanning them constantly.

Ⓐ This mayfly larva has prominent gills on its abdominal segments.

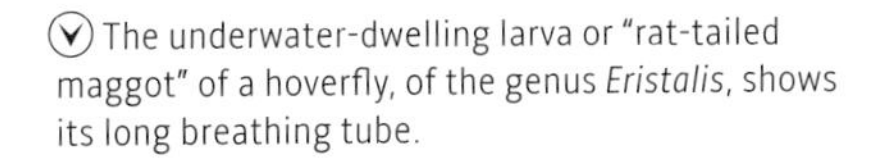

Ⓥ The underwater-dwelling larva or "rat-tailed maggot" of a hoverfly, of the genus *Eristalis*, shows its long breathing tube.

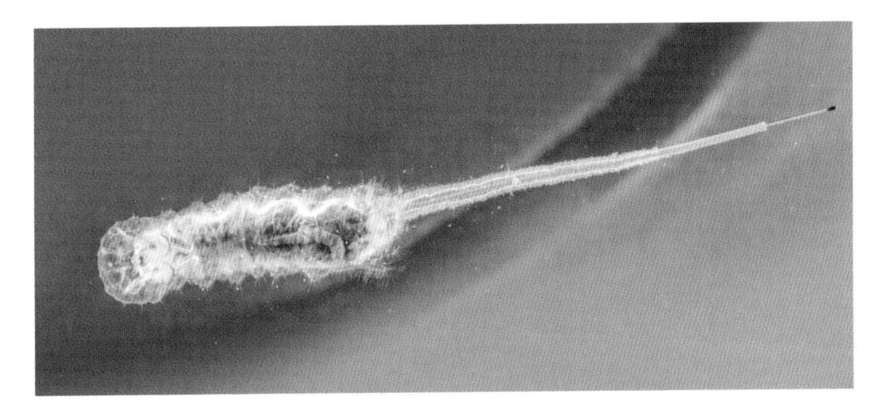

# HORMONES

**Many of the complex processes that take place in an insect's body as it grows and matures are regulated by the action of hormones, which circulate through the body via the hemolymph.**

Hormones are sometimes described as "chemical messengers." They are molecules that, when they reach their target cells, cause some physiological change to occur. For example, two hormones (prothoracicotropic hormone and ecdysone) work together to trigger the bodily changes that initiate the process of molt in larvae. The presence of juvenile hormone in larvae ensures that each molt is from one larval stage to the next. Juvenile hormone diminishes as the larva matures—when it is no longer present, the next molt is from larva to pupa. In insects that can develop in various different adult forms (for example, worker and soldier "castes" in some species of ants and termites), hormones guide them along their particular developmental pathways.

The various hormones and the body tissues that secrete them are collectively known as the endocrine system. Hormones are produced by specialized structures (glands) or specialized regions in other body parts (especially the brain), and released into the hemolymph, in due course binding to the cell membranes in their target tissues. Some hormones are proteins, others are lipids, and some are terpenes—aromatic molecules made of hydrogen and carbon chains.

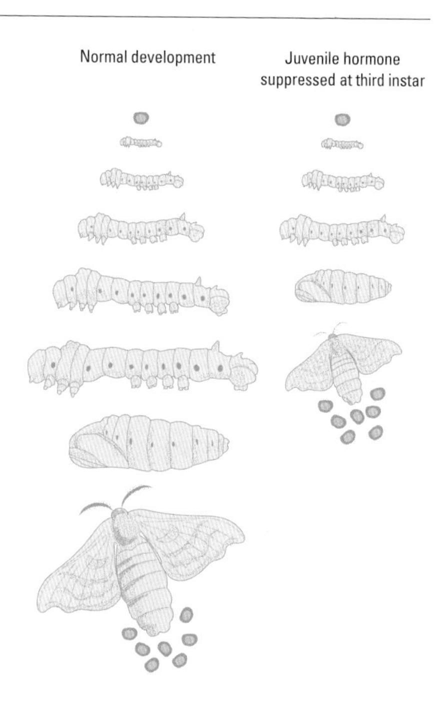


If juvenile hormone (JH) is suppressed in the caterpillar of the Silk Moth (*Bombyx mori*), at its next molt the caterpillar will pupate rather than continuing to grow, and a miniature adult will result. It is perfectly formed and, if female, will lay normal-sized eggs.

## HORMONAL ALTERATIONS

Insect development can be disrupted by interfering with hormone production. For example, if a young moth caterpillar has its juvenile hormone-secreting glands removed, it will pupate at its next molt rather than molt to its next larval stage, and from the pupa a miniature adult will emerge. Manipulation of hormones as a means of pest control is a burgeoning field of research. Certain chemicals can mimic the action of natural hormones, and cause insects to die or fail to reach maturity. As a means of pest control, hormones have the advantage of not harming non-target species (or human workers) but at present their synthesis is costly and their use is difficult (as they deactivate quickly in the environment).

Ⓐ Processes such as secreting pheromones to attract a mate are under hormonal control.

Ⓥ Hormonal activity in adult insects triggers mating behavior.

7

# REPRODUCTION

Once they reach adulthood, insects respond to hormonal instructions to go forth and breed, and for many species this is the only activity that they pursue in earnest in their brief adult lives. Different species have evolved varied anatomical and behavioral solutions to the challenge of bringing egg and sperm together.


7.1 · Male reproductive anatomy

7.2 · Female reproductive anatomy

7.3 · Mating and fertilization

7.4 · Parthenogenesis

7.5 · Laying eggs

7.6 · Unusual adaptations


Copulation can be a relatively peaceful event, but the competing reproductive needs of the sexes have also led to some remarkable anatomical and behavioral adaptations.

# MALE REPRODUCTIVE ANATOMY

**Insects have two sexes, though in some cases males are not required for reproduction. Male insects produce sperm, which they pass to the female's body when mating.**

Sexual reproduction usually involves the uniting of one each of the two kinds of gametes (sex cells)—a sperm and an ovum (egg). Males of most species produce large numbers of sperm, which are motile (freely moving) cells. As sperm cells are small and the internal apparatus to make them is also relatively small, male insects are (in most cases) smaller and lighter-bodied than females, and often with distinctly slimmer abdomens.

Sperm develops within tubules inside the testis, a small membranous pouch held in place near the rear of the body cavity by the surrounding network of tracheae. Most insects have two testes, each with a duct leading toward the outside of the body. In some more primitive species there is just one, and in some others the two maturing testes fuse together into one organ, though this still has two ducts. After exiting the testes, the ducts join into a single vessel, the ejaculatory duct, through which sperm travels to the female's body during mating. This duct may have an outpouching where sperm can be stored temporarily. On the outside of the body there is a sclerotized protrusion, equivalent to a penis, called the aedeagus, which penetrates the female to deliver the sperm. The cerci (paired appendages on the last abdominal segment) may also be used in mating, to grip the female's body.

Males of many species produce large amounts of sperm through their adult lifetimes and constantly seek mating opportunities. They are therefore often more active and travel further than females. They also spend less time feeding as they require less additional nutrition—in some species, males do not feed at all.

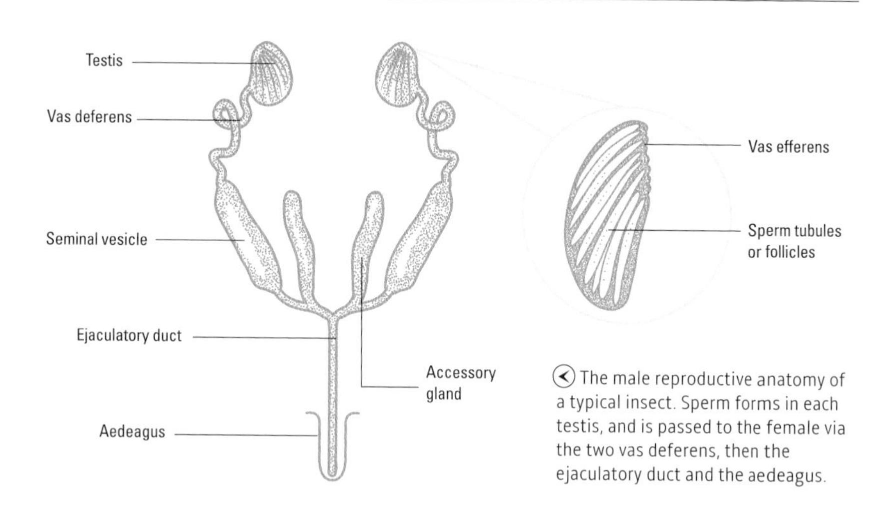


The male reproductive anatomy of a typical insect. Sperm forms in each testis, and is passed to the female via the two vas deferens, then the ejaculatory duct and the aedeagus.

## DOUBLING UP

Sperm competition is rife in some insect groups. This occurs when matings can occur freely and frequently, so males are motivated to try to ensure that their matings are successful and other males' attempts are not. In the dragonflies and damselflies, males have evolved an entire secondary set of genitalia as a sperm competition strategy. The secondary genitalia sit on the underside of the base of the abdomen, close to where it adjoins the thorax. The male transfers sperm to this structure before mating, and when he mates with a female she connects the tip of her abdomen to his secondary genitalia to receive the sperm. However, the secondary genitalia also include a tiny extensible tool that removes any previously deposited sperm from the female's body before the new sperm is added.

Ⓐ From the top—a queen, a drone, and a worker honeybee. The drones exist only to mate with queens.

Ⓥ Male Stag Beetles are much larger than females, but because the male reproductive organs are small, it is more usual for female insects to be bigger than males.

# FEMALE REPRODUCTIVE ANATOMY

**Female insects have bulkier reproductive organs than males, because forming eggs is a longer and more resource-hungry process, and elaborate anatomy may be needed to place those eggs in the right spot.**

The gametes produced by female insects, the ova, are large single cells, carrying a generous supply of yolk for the developing embryo. Because females usually produce far fewer ova than males produce sperm, they have invested more time and bodily resources in each ovum. Therefore they tend to have fewer matings and (if they can) be more selective about their partners—in some cases, they will only ever mate once.

Ova are produced by the ovaries. A female insect has two ovaries, connected to the outside world via a pair of oviducts, which join into a single tube soon after they exit the ovaries. Ova form inside tubules within the ovaries, acquiring special yolk fats and proteins that are synthesized by the fat body (see page 87). They become fertilized either in the tubules or in the oviducts, and on their journey out they acquire protective coverings, secreted by various glands.

During mating, the male's aedeagus enters the female's body and deposits sperm. In some cases, this takes place in an outpouching of the female reproductive tract called the bursa copulatrix. There are also often more than one spermatheca—these are also outpouchings of the female reproductive tract, in which the female stores sperm from recent matings to be used later. She can thus control which matings lead to fertilization and which do not.

When her fertilized eggs are ready to lay, the female deposits them via her ovipositor or egg-laying tube. This may be very small and simple, with no apparent outward

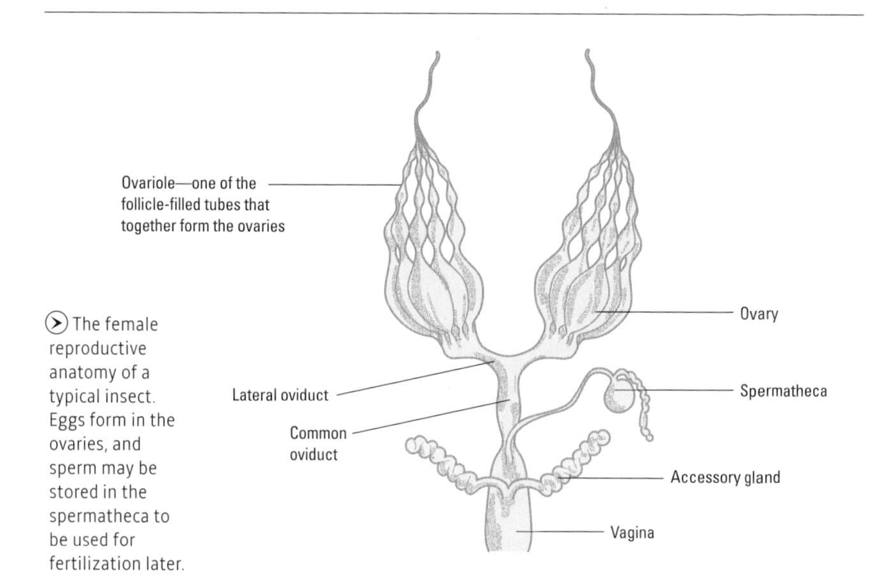


The female reproductive anatomy of a typical insect. Eggs form in the ovaries, and sperm may be stored in the spermatheca to be used for fertilization later.

Ⓐ Some dragonflies have scythe-like ovipositors, which cut holes into underwater vegetation into which eggs are laid.

Ⓐ Female ladybugs need to eat plenty of prey to provide the materials to form their large eggs.

Ⓐ Butterflies use chemoreceptors on their feet, antennae, and abdomen tips to find the right leaves on which to lay eggs.

Ⓐ Some solitary wasps have long needle-like ovipositors, to inject their eggs into beetle larvae living deep within decaying wood.

projection, or large and prominent, depending on how she lays her eggs. Bush crickets and some species of dragonfly often have long, scythe-like ovipositors, which they use to slice into vegetation, laying their eggs inside the cuts they make, while parasitoid wasps have needle-like ovipositors to stab into the bodies of their hosts. The ovipositor may be partly or fully retractable.

### STING IN THE TAIL

In bees and wasps that sting, their stings are in fact modified ovipositors—extensible, sharp-tipped, sometimes barbed, and equipped with venom glands. They may be used for defense and in some cases to kill prey. The actual process of egg-laying no longer involves the ovipositor in these species—they have developed a separate opening through which eggs are deposited.

# MATING AND FERTILIZATION

**Once an insect reaches adulthood, its priority is to breed. Finding a partner and mating successfully is not always straightforward, but the drive to reproduce overcomes all obstacles.**

When an ovum unites with a sperm, the resultant embryo carries a combination of both parents' genes. This ensures a genetically diverse population—a good thing for species survival, and most species on Earth reproduce in this way. However, for it to happen, a male and a female need to find one another and copulate. Insects locate potential partners through visual and chemical cues. Males often spend much of their active life seeking females, while females spend more time feeding, and finding suitable egg-laying sites. In some species, males perform courtship displays to demonstrate their fitness, and a female will only allow the most impressive male to mate with her. In some cases, the male may offer the female food to secure a mating opportunity.

Mating usually occurs through direct contact between the exit point of the two insects' reproductive tracts, at the abdomen tips. To achieve this, the male may climb on to the female's back, or the two may join facing away from each other. The cerci may be used to keep a grip—in dragonflies and damselflies, which mate in a more convoluted manner (see page 109), the male uses his cerci to grip the female just behind her head.

After mating, a male mantis gives his life for his future offspring—the nutrition that his body provides to his cannibalistic mate will improve the quality of the eggs she lays.

Ⓐ A female bush cricket collects a spermatophore deposited by a male.

Ⓐ Male damselflies transfer their sperm to their secondary genitalia, just behind the thorax. The female needs to bring her abdomen tip to this point to receive the sperm.

In a few insect species, including silverfish and certain grasshoppers, such as bush crickets, there is no copulation but instead the male deposits a "package" of sperm (spermatophore) and the female takes this into her body via her genital opening. The male may bring the spermatophore to his mate, or lead her to where he has deposited it.

## SPERM MEETS EGG

Each sperm and ovum contains half of its parents' genes, in a structure called the pronucleus (equivalent to the nucleus of other kinds of cells). The ovum is a large, yolk-filled cell. When a sperm swims to an ovum, it enters at a point called the micropyle to reach the ovum's pronucleus. In vertebrate animals, once a sperm cell has reached an ovum, the pronuclei of the two will fuse together, forming a single cell (zygote) with a full complement of genes. This cell then begins to divide. However, in insects, the two pronuclei remain separate, only fusing after they have each completed one cell division in their separated states, and, following this, the process of cell division is different too (see page 184).

In most insects, sex is determined by inheritance of the sex chromosome (X). Females have two copies of it and males only one, so every ovum has an X, but only 50 percent of sperm do. Sex is therefore determined by whether an ovum is fertilized by a sperm with an X (giving rise to an XX female embryo) or without (giving rise to an X male embryo).

# PARTHENOGENESIS

**Fertilization is not necessarily required for an insect ovum to develop into a larva. Parthenogenesis is the process of reproducing using unfertilized ova, and can be observed in a variety of insect species.**

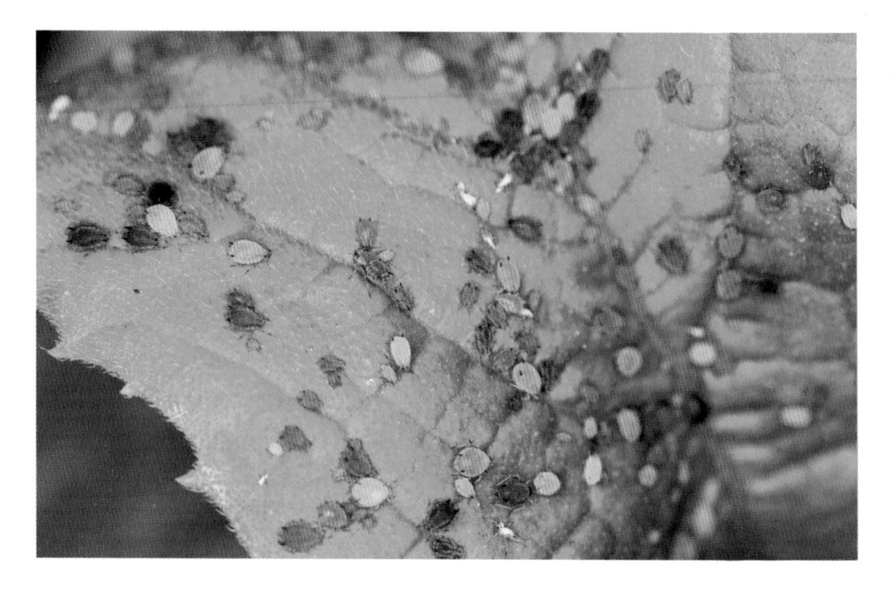

Ⓐ Aphids can build up very large populations very quickly through parthenogenesis.

Parthenogenesis is a natural and regularly occurring process in some groups of insects, most notably aphids, stick insects, and social bees, ants, and wasps. In the case of the honey bee, the colony queen mates soon after she matures and uses this stored sperm to fertilize eggs from which hatch genetically female larvae, most of them destined to be worker bees. If any are fed exclusively on royal jelly, they will become queens instead. However, she will also lay unfertilized eggs, each containing half of her chromosomes, and these will develop into male bees (drones), which are capable of mating normally with the new generation of queen honey bees.

In aphids, the mechanism is different. Wingless females give birth to genetic clones of themselves, because their ova reach maturity without going through meiosis—the special final stage of cell division that, in most cases, cuts the chromosome count in half. By reproducing in this manner, without needing to mate, a single female aphid can theoretically produce more than a billion descendants in a single season. Without a correspondingly high rate of predation, a population of aphids can quickly overwhelm its host plant (as gardeners will know). As summer ends, the females will begin to produce winged daughters, and also winged sons, also through parthenogenesis. These males have one chromosome fewer than their sisters. Winged aphids disperse from their host plant, and mate, with fertilization taking place the conventional way. The females lay their fertilized eggs on a new host plant. These eggs will hatch into wingless females the following spring, and the cycle begins again.

Ⓐ Midges of the family Cecidomyiidae are among those insect species in which paedogenesis is known to occur.

## PAEDOGENESIS

Paedogenesis, reproduction in the larval stage, involves new larvae developing from unfertilized eggs inside the bodies of female larvae. This rare form of reproduction is known in a few species, including some forms of midges and beetles. The larvae will often consume their mother's body, and the same process may be repeated through several generations of larvae before any individuals actually reach pupation stage.

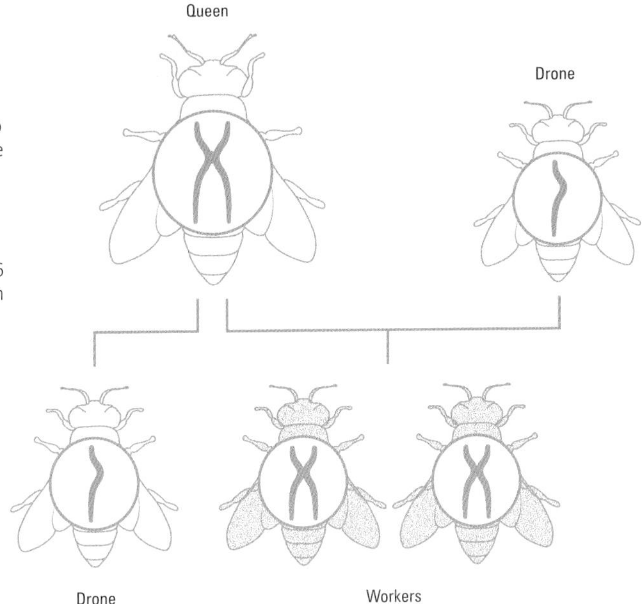


Ⓐ A queen honey bee can produce drone (male) offspring without mating, but to produce daughters she needs to mate with a drone. Females (workers and queens) are diploid (they have a pair of each of the 16 chromosomes, so 32 in all) while drones are haploid (just one of each chromosome, so 16 in all).

# COURTSHIP AND MATING BEHAVIOR

*The drive to reproduce has given rise to some of the most peculiar and complex of all insect behaviors, involving competition, coercion, and, in a few cases, harmonious cooperation.*

Ⓐ A pair of dung beetles work cooperatively to make the dung ball on which the female will eventually lay her eggs.

The way that animals choose potential mates has a lot to do with what level of parental care they provide to their young. For animals that work together to rear their young, courtship is often elaborate and prolonged, with both sexes playing an equal part, each assessing the other's potential parenting skill. However, in many species the young receive care only from their mother, and in many more (including nearly all insects), there is no care at all once the eggs have been laid. In these cases, courtship behavior tends to be limited to males showing off their health and fitness to females, while fighting away other males.

Some male insects, including many butterflies and dragonflies, establish and defend a territory from other males. They may oversee the territory from a high perch or patrol it on the wing. Any male that encroaches will be chased off so he doesn't get in the way when a female approaches. Male butterflies often have specialized scent-producing scales on their wings (androconia), and when they approach a female, they shower her with this scent by walking or flying close to her, vibrating their wings. A stylized chase may ensue—for example, the male Silver-Washed Fritillary (*Argynnis paphia*)

(v) Antennae contact is part of the courtship ritual in many insect species.

(v) Male Orchid Bees collect fragrances from orchids, though it is not yet known whether they use the scent as part of their courtship display.

loops around the female as she flies in a straight line.

An elaboration of territorial behavior is lekking, in which males make small territories (leks) close together, where they perform courtship displays within sight and sound of each other. The Orchid Bee (*Eulaema meriana*) of Central and South America, forms leks in which males display by buzzing their wings and showing off yellow markings on their abdomens. Females watch the males performing and choose the one they find most impressive as their mating partner.

## Pair bonds

It is very rare for male and female insects to establish a long-lasting, monogamous bond. Exceptions include dung beetles, in which pairs work together to create a dung ball that will feed their young. The queen and king in a termite colony may also remain together for many years. Pairs of the endangered Lord Howe Island Stick Insect (*Dryococelus australis*) are said to form bonds, preferring to mate with their regular partner.

# LAYING EGGS

**While the insect male's involvement in breeding is usually over after mating, the female has another vital task to perform—she must lay her eggs, in the right place for the larvae to thrive.**

Not all insects carefully place their eggs on a suitable substrate. Some butterflies, such as the Ringlet (*Aphantopus hyperantus*), let their eggs fall freely. At the other end of the parental care scale, some solitary bees and wasps lay their eggs in burrows or even elaborate structures built from clay. Certain solitary bees only lay their eggs inside old snail shells, creating a nest chamber at the center of the shell's inner spiral and blocking the rest of it with tiny stones to protect the nest from predators. Dung beetles work together in pairs to make and bury a ball of dung—a food supply for the larvae, which hatch from eggs laid on the dung.

Eggs may be laid individually or in clumps—in plant-eating species such as shield bugs, and many Lepidoptera, a large and tightly packed egg clump is laid on the food plant and the larvae will stay close together while they are very small. Insects with active, predatory larvae are often less choosy about egg-laying sites and also tend to lay eggs individually (not least to reduce the risk of cannibalism).

If the egg needs to be attached firmly to its substrate, it emerges with a sticky coating. With lacewing (family Chrysopidae) eggs, the coating sticks to the substrate, and extends as a thin stalk that hardens in the air as the female raises her abdomen, with the egg ending up stuck at its tip. The stalk holds

Ⓥ Ringlets are among the few butterfly species that do not lay their eggs directly onto a plant.

Mason bees (genus Osmia) block the entrances to their nests with debris once they have laid their eggs. Here, a new adult bee digs its way out.

Shield bug eggs are laid in clusters and the young larvae will live communally after they hatch.

the egg clear of the substrate, making it less vulnerable to opportunistic predators. Bush crickets and many sawflies hide their eggs in a slit cut into the food plant, made using the sharp ovipositor. Many dragonflies and damselflies also have cutting ovipositors which slice into underwater vegetation, making a shelter in which an egg can be laid. Damselflies of the genus *Lestes* may lay their eggs in plants above water, leaving visible signs of their activities in the form of rows of scars on the stems.

Blowflies lay their clusters of soft-shelled eggs directly onto carrion. The eggs hatch very soon after laying, as there is no time to waste—fresh meat decays quickly as bacteria start to colonize, and in the meantime there is fierce competition between different kinds of scavenging animals for this scarce resource.

**LIVE YOUNG**

While most insects lay eggs, a few give birth to larvae. Sometimes this is because normal, yolk-filled eggs form and hatch inside the female's genital tract. In the case of tsetse flies (genus *Glossina*), the hatched young remain inside the female for their entire larval life span, consuming food stores that she holds in her body. They only emerge when fully grown and ready to pupate. Aphids, which do not have a pupal stage, are born live as miniature replicas of their wingless mothers. Insects of the order Strepsiptera, which spend most of their lives as internal parasites of other insects, do not form eggs at all but free embryos, without any kind of binding shell or membrane around them. They obtain nutrients from their mother's body through osmosis until they are born as larvae capable of independent life.

# UNUSUAL ADAPTATIONS

**With their brief adult life spans, insects need to be able to breed quickly and efficiently, and this need has led to the evolution of some remarkable anatomies and behaviors.**

Broadly speaking, in terms of mating, the adult male insect's goal is to father as many offspring as possible, while the female's is to choose the "best" mates to fertilize her supply of eggs. These goals are not necessarily compatible. One way males try to prevent female choice and ensure their own paternity is the use of mating plugs—placing a physical obstacle in the female's genital tract so she cannot mate again. This is seen in some bees and butterflies—the sperm of the male Rocky Mountain Parnassian Butterfly (*Parnassius smintheus*) is bound up in a waxy plug that also releases chemicals that discourage other males from approaching the female (though the plug does also include nutrients that help her to form her eggs).

In mantids, a male can further his chances of fathering many young by making the ultimate sacrifice and being literally consumed by his partner. Females lay more eggs if they eat the male after mating than if they do not.

While sperm cells are usually very small and produced in great abundance, fruit flies of the genus *Drosophila* produce far fewer, and much bigger sperm. *Drosophila bifurca* produces sperm that, at full length, measure more than 2 inches (5cm)—the longest sperm cells of any animal. It is thought that

Ⓥ Male *Parnassius* butterflies have wax-secreting glands in their reproductive tracts to produce a mating plug.

female choice has, over many generations, favored males producing larger sperm, perhaps because it indicated that they were healthier and so were able to dedicate more bodily resources to building these "super sperm."

The female parasitoid wasp *Copidosoma floridanum* lays just one or two eggs inside her caterpillar host's body, but each egg will give rise to 2,000 to 3,000 embryos. These are genetic clones, each forming from the single initial zygote. More extraordinary still, some of these develop into specialized, large-jawed larvae that will never become adults. Instead, they lead an active life as "soldiers," roaming the host's body and killing any unrelated parasitoid larvae they encounter, thus removing competition for their siblings. The soldiers die when the fully grown normal larvae emerge from the host to pupate. The developmental pathway each embryo takes is determined by whether it forms a primary germ cell during cell division—those without these cells will become soldiers.

Ⓥ Fruit flies produce unusually large and long-tailed sperm.

Ⓐ Female mantises have special glands that exude a dense, foamy material. They secrete this foam over their egg clusters to provide protection from predators.

## THE LOSS OF CHOICE

The evolutionary struggle of male animals to overcome females' ability to choose their mates is exemplified in bedbugs (family Cimicidae). Male bedbugs bypass the female's genital tract completely and instead pierce a hole in her abdomen cuticle, and release sperm into a storage area in her body cavity (the spermalege). This is known as "traumatic insemination," though the spermalege apparently helps to heal the female's wound. If a male bedbug inseminates another male (as sometimes occurs with unusually large males, which other males mistake for females), the inseminated male is badly injured and likely to die.

# INVASIVE INSECTS

*An invasive species is classed as one that is introduced to a region outside its native lands, where it spreads widely and proves damaging to ecosystems in some way.*

Most nations on Earth have experienced the problems caused by invasive species of one kind or another. In the UK, the Eastern Gray Squirrel from North America has largely eliminated the native Eurasian Red Squirrel, while in North America native hole-nesting bird species such as bluebirds are threatened by the House Sparrow and Common Starling from Europe. Australia's long list of invasive species includes the rabbit, the Cane Toad, and even the honeybee, which escaped from managed hives to form feral colonies and now outcompetes native mammals and birds for nesting sites.

When a species is introduced to a new area, it is likely to come into competition with native species that occupy the same natural niche. Those species that become invasive are the ones that outcompete and even eliminate their rivals with ease. The Asian Lady Beetle (*Harmonia axyridis*) was intentionally introduced to North America in the 1970s to serve as a pest controller. But while it does prey on crop-destroying aphids, it also feeds on the eggs and larvae of other ladybugs, and its arrival led to a decline in many of North America's native ladybugs.

Several ant species have become invasive in various parts of the world. Argentine Ants (*Linepithema humile*) have become established in many regions of the world and tend to displace other ant species with ease. Various species of invasive ants are voracious predators and can harm many other animals, including vertebrates. They may also damage forests, through guarding

Ⓥ The Asian Lady Beetle, accidentally introduced to North America, consumes aphids but also preys heavily on other ladybugs.

Ⓐ Lily Leaf Beetles will damage large ornamental irises as well as lilies and fritillaries.

Ⓐ The Argentine Ant is a devastating invasive species now found in many parts of the world.

aphids and scale insects.

The introduction of insects to new regions is usually accidental—often they have been transported as eggs or larvae inside fruit, vegetables, or shipments of lumber and wood furniture. For this reason, some countries, such as New Zealand, have rigorous biosecurity rules. It is not permitted for visitors from overseas to bring any fresh fruit, vegetables, or animal products into the country. Wherever you live in the world, if you find a live, unfamiliar insect inside packaged food or another plant product, it is important you do not release it into the wild. Instead, it is best to contain the insect and contact a local animal charity or natural history museum for advice.

## Host-specific hazards

Many herbivorous insects are specialist feeders on one plant species, and when they are introduced to an area where their host plant occurs, but their natural enemies (such as parasitoid wasps and flies) do not, they can do devastating harm to their host plant. Examples include the Box Tree Moth (*Cydalima perspectalis*) which causes severe defoliation of box trees and has been introduced to western Europe from the far east of Asia, and the Lily Leaf Beetle (*Lilioceris lilii*), which feeds on members of the Liliaceae family, and has been introduced to North America from southern Europe and Asia.

Ⓐ Strict biocontrol measures are needed at docks and ports to prevent further accidental spread of invasive species.

# INSECTS AS CROP PESTS

*Ever since humankind began cultivating plants to harvest, we have been doing battle with a range of insects that have a taste for the same plants as ourselves.*

When we set out to grow a crop of a particular plant, whether in our back yard or on a wide expanse of farmland, we are creating an unnatural habitat in the outside world. Monocultures rarely occur in nature—only where conditions are so hostile that just a few highly adapted specialists can survive. Elsewhere, plant communities are naturally mixed, and the animals that depend on those plants tend to make use of many of them. Those herbivores that feed only on one species of plant will normally have scattered populations, as individuals of their host plant will be scattered.

Creating a habitat occupied by just one plant species can encourage unusually large populations of particular insects, sometimes to the extent that they can destroy the entire harvest. A famous example is the Colorado Potato Beetle (*Leptinotarsa decemlineata*). These beetles and their larvae are not actually interested in the potatoes, but consume the plants' foliage, drastically reducing yield of potatoes below ground. Crop pests can also harm plants we grow for nonfood purposes, such as the Cotton Aphid (*Aphis gossypii*), which feeds on the stems and leaves of cotton plants (and also attacks other crop plants including coffee, cocoa, citrus trees, and cucumbers).

Locusts are perhaps the most notorious of all crop-eating insects. When their numbers rise enough for them to enter a swarming phase, they can cover huge distances and entirely devastate all kinds of food crops over a wide area. Locust "plagues" occur at intervals of several years, and when they do, they can cause famine, which may result in mass human migration. Careful monitoring of their numbers and decisive action when an outbreak occurs is necessary to preempt these devastating events, as occurred in western Africa in 2004 when over 50,000 square miles (around 130,000 sq km) were treated with insecticide to control an upsurge. This action stopped the upsurge, but crop losses valued at $2.5 billion had already occurred.

Ⓐ These corn stems have been damaged by larvae of the widely occurring moth *Chilo partellus*.

Ⓑ The Colorado Potato Beetle, native to the western US and Mexico, is now present throughout North America and much of Eurasia.

## Counting the cost

Crop pests' depredations can harm regional economies, and controlling crop pests is expensive too. Small-scale farmers across six eastern African countries lose an estimated $450 million worth of their maize crops each year to just one pest species, the Spotted Stem Borer (*Chilo partellus*), a moth that was accidentally introduced from Asia.

The battle against crop pests sometimes involves wholesale use of insecticides, but there is an ever-growing research effort into finding new, better ways to control specific problem species, without affecting other insects (many of which are allies in our battles against the pests). Bioengineering species-specific viruses, bacteria, and other pathogens is one productive field of research.

Ⓥ Crop-spraying with insecticides is effective against pests but may harm many other non-target species.

Ⓥ Aphids can be very damaging, partly because they can build large colonies so rapidly.

# EGGS AND LARVAE

For most insects, life begins as an egg, placed in a safe and food-rich place. Within the egg, a larva develops, and once it hatches, its objective in life is to eat and to grow. Larvae can be very different in both appearance and habits to their adult selves.


8.1 • Types of eggs

8.2 • Development in the egg

8.3 • Types of larvae

8.4 • Feeding

8.5 • Growth and molt

8.6 • Lifestyle changes


Many hazards threaten insects in their early life stages. Sawfly larvae bunch closely together for safety in numbers.

# TYPES OF EGGS

**An egg must nourish and protect the developing embryo inside, until it has become a fully formed larva and is ready to hatch. To achieve this, eggs have evolved in varied forms.**

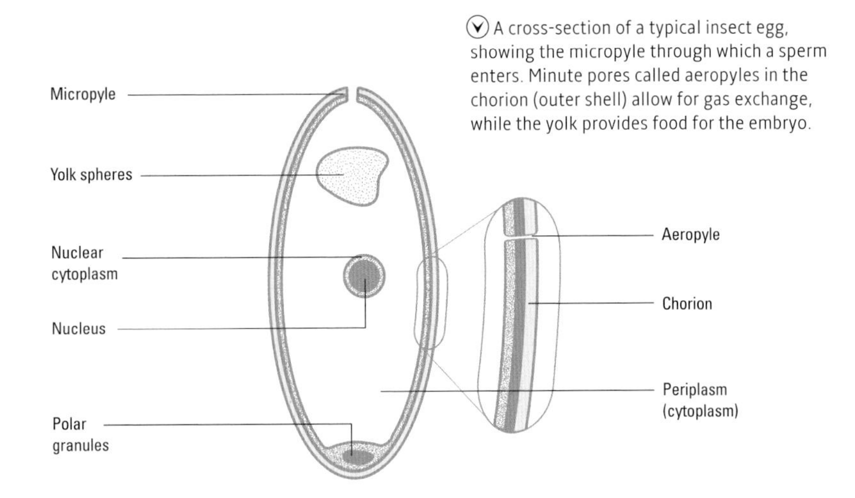


A cross-section of a typical insect egg, showing the micropyle through which a sperm enters. Minute pores called aeropyles in the chorion (outer shell) allow for gas exchange, while the yolk provides food for the embryo.

Insect eggs contain a developing embryo (which is sometimes an almost fully developed larva by the time the egg is laid), and a supply of yolk for the embryo to consume. Their outer shell (chorion) contains small perforations called aeropyles to supply the embryo with air. Some eggs have a very soft chorion while in others it is more brittle. The micropyle, where sperm entered the egg, is an often visible depression.

Eggs are very vulnerable to predators, being small and immobile. Some insects protect their eggs by laying them in a shelter or nest, but many more are laid in the open. Those that must survive in the outside world for long periods (for example, over winter in temperate countries) are particularly at risk and have special adaptations to protect them. The eggs of the Brown Hairstreak (*Thecla betulae*) a butterfly of northern Europe, for example, are flattened disks with thickened, spiky shells.

Phasmid eggs often closely resemble seeds, in both appearance and scent, and are

The microstructure of some butterfly eggs is extraordinarily intricate.

mistaken as such by ants, which bury the eggs in their nests as food stores. This protects the eggs from other predators and they hatch safely underground. Those phasmid eggs that are eaten by birds can sometimes survive the digestive process, emerging from the bird's body intact, and the insects can disperse to new habitats in this way.

Phasmid eggs show remarkable variety and mimic different kinds of large seeds.

## UNDERWATER EGGS

Eggs laid in water may need special adaptations. Those of some mayflies are simple and unprotected, but the free-swimming larvae hatch out within a minute or two of being laid. Others that will spend longer in water may have a trapped air layer (a plastron) for breathing, or an extension of the chorion, which reaches into the air as a breathing tube. Eggs laid underwater often also have a thick coating of gel-like material, which protects them from drying out if water levels fall and leave them exposed to the air.

The eggs of the Brown Hairstreak Butterfly need to survive the winter with a tough casing and a flattened shape.

# DEVELOPMENT IN THE EGG

**Inside the egg, the embryo must grow from a single cell to a mobile larva that is capable of breaking through the shell and then beginning its active life outside the egg.**

Once the sperm and ovum's nuclei have fused and a zygote forms, its nucleus begins to divide. There is still just one cell at this stage, but the number of identical nuclei it contains increases rapidly. When several thousand nuclei have formed, they move through the egg to form a layer around the outside, containing the yolk within. Here, each nucleus forms a membrane around itself and becomes a fully formed cell.

Now, the cells continue to divide and begin to differentiate into different types of tissues. Those on one side of the egg become a distinct region (the germ band) that starts to form the body tissues, while the rest form a membrane (serosa) that encloses the embryo and yolk together. The embryonic cells absorb the water and nutrients they need directly from the yolk, and waste carbon dioxide diffuses out through the chorion's aeropyles.

The germ band begins as a flat sheet but then folds in on itself. The outer layer will become the larva's cuticle, as well as its brain and nervous system and the initial and terminal stretches of the digestive tract (with their cuticle lining), while the inside layer forms the circulatory system, internal sex organs, and muscles. A third, innermost layer develops later and becomes the central part of the digestive tract, the section that lacks a cuticle lining. As it develops, the tubular germ band takes on a segmented appearance and the larval appendages begin to show as tiny buds on the outer side. As it grows, the

Ⓥ The strong biting mandibles of a butterfly caterpillar make short work of its own egg shell.

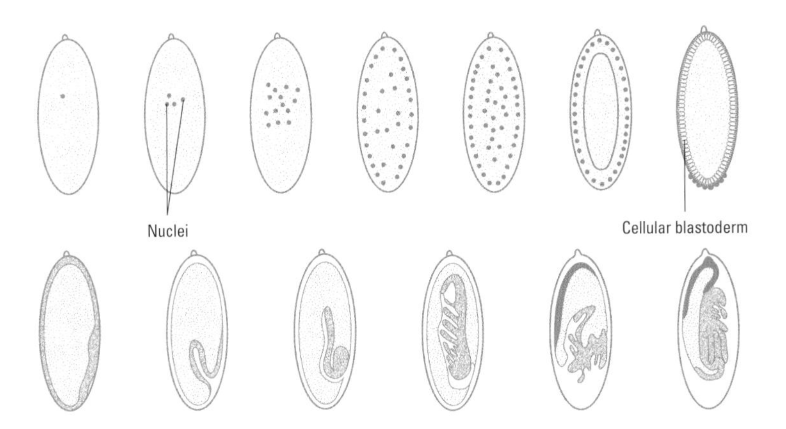


Ⓐ In a developing insect egg, the original nucleus divides several times, then these nuclei migrate to the outer edge of the egg, where they form individual cells. The originally flat layer subsequently produces a distinct fold and this is where the embryo will develop.

⊙ Close to hatching, the insect eggshell often becomes translucent and the larva is visible inside.

embryo gradually consumes the supply of yolk inside the egg—this is its food and water supply.

## HATCHING

When the larva is fully developed and ready to hatch, it may emerge by chewing a hole in the chorion, or it may take in extra air through the chorion's aeropyles, to become large enough to rupture the chorion with its expanding body. The broken chorion is, in some cases, the first meal that the newly hatched larva will consume. Larvae of some species will eat any other eggs of their own kind that they encounter. In species with this cannibalistic tendency, the female will not usually lay more than one egg in the same place, but if she is injured she may be forced to lay several together—as a consequence, the first one to hatch may well destroy all the rest.

# TYPES OF LARVAE

**From a squirming, blind, and legless maggot to a dragonfly larva that swims, stalks, and strikes out in its pursuit of prey, insects in their larval form are hugely varied.**

The larvae of insects that undergo incomplete metamorphosis (that is, they molt directly to their adult form, rather than passing through a pupa stage) tend to have a far more developed and adultlike appearance right from the moment they hatch. Such larvae are sometimes known as nymphs, and they are typically quite mobile, with well-developed legs and sensory apparatus. Larvae of grasshoppers, crickets, and others in the order Orthoptera are especially adultlike, but species that metamorphose incompletely and have an aquatic larval stage, such as mayflies and dragonflies, look quite different, having adaptations to their underwater lifestyle that are lost in adulthood, such as their gills.

Larvae of true flies that feed on carrion or dung are among the least developed, as they hatch from eggs laid directly in the midst of enough food to sustain them until pupation. They have tubular bodies with no legs or eyes, a thin body cuticle and tearing, hooklike mouthparts. Wasp, bee, and ant larvae, which develop in the shelter of nests built and stocked with food by their mothers, are similar in appearance, although the related sawflies, which feed on foliage, are more active with patterned bodies and six well-developed climbing legs, and more closely resemble butterfly and moth caterpillars.

Beetle larvae are highly variable. Those that develop within decaying wood look rather maggot-like but nonetheless have six obvious legs and more developed, biting mouthparts. Larvae of diving beetles are often very strong, active swimmers and hunters. Ladybug larvae are strong-legged, tough-bodied active predators that dwell on

A lacewing larva under its casing looks like a clump of lifeless organic matter.

Ⓐ A bagworm (the caterpillar of a moth from the family Psychidae) inside its protective case.

Ⓛ Larvae of *Perreyia* sawflies forage on the ground in close-knit groups, which are difficult for predators to attack.

tough-bodied active predators that dwell on foliage, whereas some herbivorous beetle larvae are slow-moving and slug-like with tiny legs and thin cuticles.

## HOMEMADE SHELTERS

Growing up inside a protective shelter helps reduce the risk of predation, and some larvae build their own. Caddisfly larvae build a portable case around their bodies from tiny bits of gravel, twigs, and other materials they find in the water, held together with silk from their mouth glands. A few species of land-dwelling moth caterpillars build cases too—the so-called "bagworms" (family Psychidae), whose cases can be 6 inches (15cm) long in some species. The larva of the Lily Leaf Beetle (*Lilioceris lilii*) covers its body with heaps of its own frass (excrement) to help protect it from predators, while lacewing larvae may place the remains of their prey on their backs as a means of disguise.

# FEEDING

**As the children's book *The Very Hungry Caterpillar* explains, eating is the main activity of all larval insects. During the most active growing phase of their life cycle, larvae consume almost anything, and feed in different ways.**

Although some adult features are still entirely undeveloped, such as wings and associated musculature, and reproductive organs, insect larvae have the same basic feeding apparatus as adults—a set of mouthparts with salivary glands, a foregut, midgut, and hindgut. Their ability to consume food quickly and grow rapidly is well known. Moth caterpillars may increase their weight ten-thousandfold between hatching and pupation. Some common species that live colonially, such as the Gypsy Moth (*Lymantria dispar*), are classed as pests because they can rapidly defoliate large numbers of plants, harming crops and wild habitats.

The very tiny larvae of leaf-mining moths live inside plant leaves, eating only the inner cells. Their activity shows as pale tracks or patches through the leaf's intact outer layer (cuticle). Many types of plants have defenses against plant-eating insects, in the form of noxious chemicals, but insects have evolved various ways to circumvent these, from feeding at certain times of day when toxin production is low to producing chemicals in their guts that neutralize the toxins.

The predatory larvae of larger diving beetles are unusual in that, as they near their maximum size, they frequently prey on vertebrate animals, such as small fish and amphibians—their mandibles are adapted to seize and puncture the prey, and then they pump digestive juices into the wound, via a groove in their mandibles, and suck up the resultant predigested liquid.

Gypsy Moth caterpillars feed in large groups and can defoliate entire trees.

Ⓐ The pupa of a social wasp. Adult wasps nurture larvae within the nest, feeding them on insects that they cut up into bite-sized pieces.

Some insect larvae are detritivores, feeding on dead and decaying organic matter. Caddisfly larvae eat this kind of material, searching for it on the bottom of the lake or pond where they dwell. Larval fleas consume the droppings of adult fleas (which have a high content of dried blood).

**FEEDING THE BABIES**

Social insects—bees, ants, wasps, and termites—prepare food for the larvae in the colony. Larval honey bees are fed on a mixture of protein-rich pollen and nectar, delivered to them by workers. In the case of social wasps, the workers dismember prey that they catch (mainly other insects) or cut up pieces of carrion, which they feed to the larvae. In return, the larvae produce a sugary liquid, which the worker wasps eat.

Some of the other insects that provide parental care will give food directly to their young. An earwig mother brings back tiny fragments of organic matter to the nest and feeds it to her larvae, and she will also give them regurgitated food. Pacific Beetle Cockroach (*Diplotera puntuata*) mothers nourish their young on a unique protein-rich bodily secretion (see page 162 for more details).

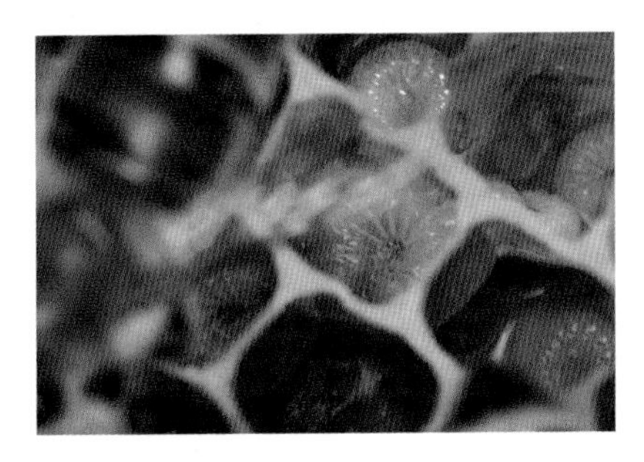

Ⓐ Honey bee larvae are fed by adults on "bee-bread" (pollen and nectar mixed together).

Ⓐ A termite worker searches for decaying plant matter, which it will swallow and later regurgitate for the colony's larvae.

# GROWTH AND MOLT

**The relatively rigid insect exoskeleton offers protection, but poses a problem for larvae, which need to grow. The solution is a series of molts—all insect larvae need to molt to grow rapidly.**

The stages of a larva are known as instars, or sometimes stadia. A newly hatched larva is called a first-instar, and after its first molt it is a second-instar, and so on. The number of molts a larva goes through ranges from just four or five to 15 or more, and in some species the number of instars males and females undergo is different. In insects that metamorphose incompletely, wing buds appear in later instars, within which the wings grow, ready to unfold in the final metamorphosis into the full adult form.

As we have seen before, the timing of molt is under hormonal control, and the volume and movement of hemolymph plays a role in allowing the larval cuticle to break (see pages 100–01). The larva will also encourage the process, arching and pushing its body against the point where the cuticle fractures. When a butterfly caterpillar molts, its cuticle breaks behind the head, and the head capsule detaches first. The caterpillar then wriggles forward out of its old cuticle, which wrinkles up like a sock. With dragonfly larvae, the cuticle fractures along the back, and the larva eases its body, head, and legs out through this opening, leaving behind the discarded cuticle as a tissue paper-thin but almost intact, rigid shell (exuvia).

Instars of the Monarch Butterfly caterpillar, showing the prodigious rate of growth over just a couple of weeks.

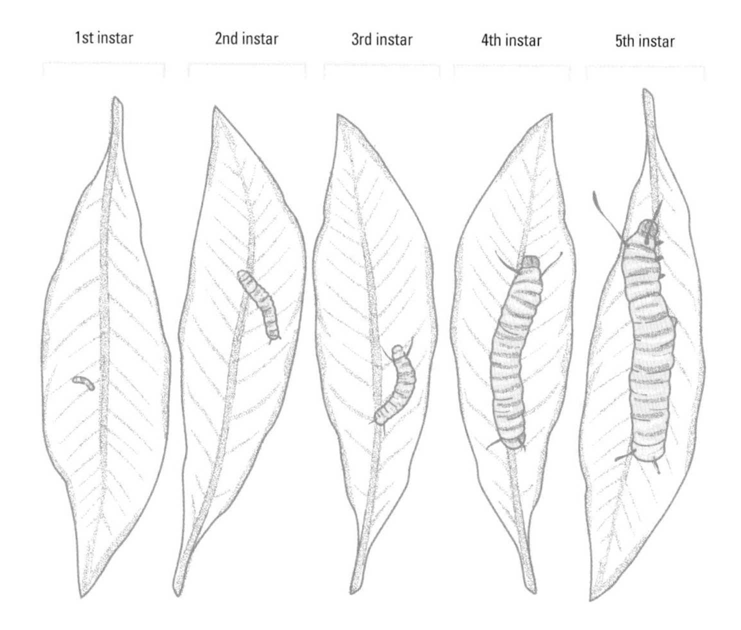

Ⓐ The cast-off exoskeleton of a grasshopper larva following a molt.

Freshly molted larvae have soft, rather than brittle, cuticles. They often look pale, with muted colors and patterns, over the hours or days that it takes for the cuticle to expand and then harden. This is a vulnerable time, during which larvae may be less active, to reduce their chance of being spotted by a predator. Some species may eat the exuvia immediately following the molt.

#### NEW CLOTHES

Some insect larvae change little in their outward appearance from one instar to the next, apart from growing larger, while others change markedly. In its early stages, the caterpillar of the Swallowtail Butterfly (*Papilio machaon*) is black with a whitish patch and resembles bird droppings. However, its final instar is white with bold black and orange spotting—effective camouflage at a distance, but at close range this pattern functions as conspicuous warning coloration, indicating (truthfully) that it tastes unpleasant.

The shield bugs or stink bugs (infraorder Pentatomomorpha) are another group in which appearance changes considerably through the various instars. Early-instar stink bugs tend to have a round, domed form before they develop the broad-shouldered, shield-shape of adults. They are also often more strikingly colored and patterned than adults—these insects tend to spend winter hibernating in their adult form and so need effective camouflage to help keep them safe while they are inactive.

# LIFESTYLE CHANGES

**As they mature and grow, and eventually reach adulthood, insect larvae of some species go through quite dramatic changes in behavior, and in their ecological role.**

We know that many insects, especially those that fully metamorphose, change their feeding methods—and sometimes their diets—completely from larva to adult. Many others switch from a fully aquatic life to being terrestrial. Some insects are highly gregarious as larvae but solitary as adults, while with others the opposite is the case. Metamorphosis is far more than a complete transformation in appearance.

Changes that occur through the nymphal stages, before reaching adulthood, can also be dramatic. Butterfly caterpillars of the genus *Phengaris* begin their first four or so instars feeding on plant foliage, but live inside ants' nests and become carnivores in their later stages (see page 172).

Locust larvae change their behavior and appearance according to their population density. From their early instars, they follow a developmental pathway into either a solitary or a gregarious (swarming) form, depending on numbers. This is governed by the neurotransmitter serotonin, which is released in a larva when it receives high levels of tactile stimulation from other locust larvae. The swarming form is the result of overcrowded conditions. It feeds more avidly, travels further and faster, and (when adult) reproduces more quickly than the solitary form, and also has a different appearance. Locust swarms only occur periodically, when populations of solitary locusts have built up to sufficiently high levels to trigger new

Large Blue Butterfly caterpillars (genus *Phengaris*) spend most of their lives inside ant nests.

larvae to become the gregarious form. By swarming, locusts may be able to reach and colonize new areas, but the cost in human terms can be serious—swarms feed voraciously and can entirely devastate crops over wide areas.

Biochemical changes caused by overcrowding will trigger locust populations to swarm.

## BEAST TO BEAUTY

The antlions (family Myrmeleontidae), in their adult form, are large but delicately beautiful insects with long slender bodies and two pairs of broad, patterned wings. Most feed on nectar and pollen. As larvae, they are ambush predators par excellence. Using their flattened, squat abdomens as shovels, they dig pitfall traps in soft sandy ground—funnel-shaped depressions with steep sides. When an ant or other insect slips down the bank of the trap, the antlion larva emerges rapidly from the center of the pit and seizes the prey in its huge mandibles.

The mantidflies of the subfamily Mantispinae are relatives of the antlions and also pursue rather different (though equally predatory) ways of life as larvae and adults. The larvae climb onto female spiders and stay with them until they lay eggs. They feed on the eggs and pupate inside the egg-sacs. The large-winged adults seize smaller insect prey with their mantis-like barbed front legs.

The antlion larva's ingenious pitfall trap (center) and an adult antlion (right).

# METAMORPHOSIS

Ever since we realized that the caterpillar was the infant form of the butterfly, we have been fascinated by insect metamorphosis. It is difficult to think of any other natural transformation that is so dramatic, especially given the short time frame of an insect's life. We still do not fully understand the physiological changes involved.


9.1 • Types of life cycle

9.2 • Incomplete metamorphosis

9.3 • Full metamorphosis

9.4 • Transformation within the pupa

9.5 • Emergence

9.6 • Maturation in adulthood


⊳ A cicada struggles free of its larval casing. Most insects will undergo a similarly marvelous transformation to reach their adult form.

# TYPES OF LIFE CYCLE

**Insects are not the only animals that change their form through their lives, but for human observers they are the most dramatic and easily observed shape-shifters.**

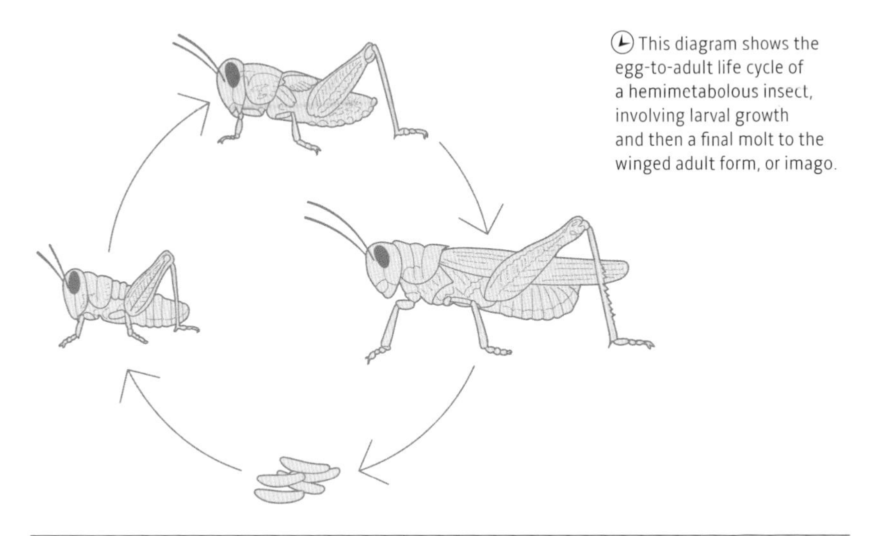

This diagram shows the egg-to-adult life cycle of a hemimetabolous insect, involving larval growth and then a final molt to the winged adult form, or imago.

Most insects begin life within an egg, and most have a nonreproductive growth phase, with regular molts to allow this growth, before reaching adulthood. This is the point where growth ceases and the insect becomes capable of breeding; in most cases, it also becomes winged.

The insect species that have changed the least from the world's earliest examples of insectkind show relatively little change in appearance as they grow. The silverfish, for example, resemble miniature adults from the moment they hatch out from their eggs. They become larger with each molt but never develop wings. They also continue to grow (very slowly) and molt regularly after becoming reproductively mature.

Insects that are winged as adults but do not pass through a pupal stage are known as hemimetabolous, meaning that they undergo an incomplete metamorphosis. They include the grasshoppers, crickets, true bugs, dragonflies, mantids, mayflies, and stoneflies. Their adult form may be very similar in appearance to the larval form that preceded it. Larvae of hemimetabolous insects are active with well-developed legs and are easily mistaken for adult insects. Their lifestyle and diet are usually similar if not exactly the same as the adult's, though in the case of species that live underwater as larvae, the transformation is more dramatic.

The insect groups that have diverged most dramatically from their ancestors are the holometabolous species—those that pass through a mostly inactive pupal stage during which their bodies are remodeled from their larval form. They include the butterflies and moths, bees, wasps, and ants, true flies, and beetles. Larvae of holometabolous insects are,

Silverfish have no distinct adult stage—their growth pattern is sometimes known as ametabolous metamorphosis.

in general, less active, softer-bodied, and "vermiform" (wormlike) compared to those of hemimetabolous insects.

**EVOLUTION**

Other arthropod animals also go through a series of molts as they mature, but the development of a pupal stage has only evolved in insects, and first occurred some 280 million years ago. Much debate has raged through the last couple of centuries over how complete metamorphosis evolved, and the answer is still not fully clear, but it is likely that holometabolous insects evolved from hemimetabolous ones that began to hatch from their eggs at an earlier, more embryonic stage of development. Some hemimetabolous insects today, such as damselflies, do hatch in a much less developed and more vermiform state—sometimes known as a prolarva—which quickly molts to a more adultlike second-instar larva. Given the right conditions, such undeveloped larvae can make use of different resources from the adult insects. This allows the species as a whole to use a broader niche. The majority of modern insect species on Earth are holometabolous.

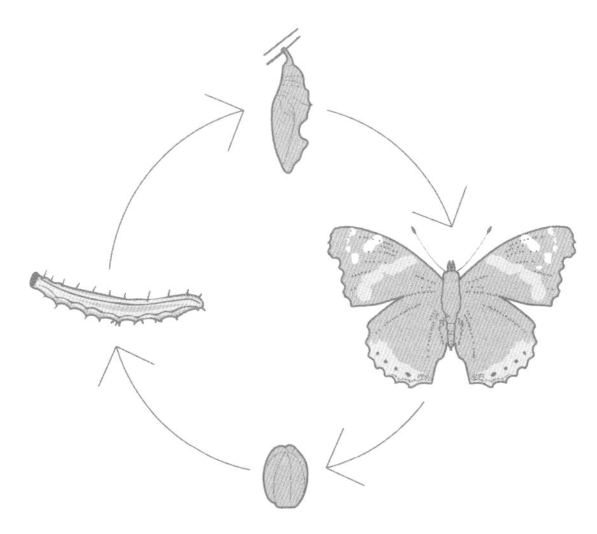

This diagram shows the four-stage life cycle of a holometabolous insect: from egg, to larva, to pupa, to winged adult (or imago).

# INSECT LIFE SPANS AND LIFE CYCLES

*It is often said that a mayfly only lives for one day, but that is only true if we disregard the preadult stages of its life. Many insects are in fact surprisingly long lived.*

The periodical cicadas (genus *Magicicada*) of North America live much longer than most small mammals. However, virtually all of that life is spent as a larva, living underground and sucking sap from tree roots. On a warm night in their 17th spring (or 13th in some species), they finally emerge and molt to their adult form. They live for no more than six weeks as adults. The synchronous emergence of each population helps ensure that all individuals can easily find a mate, and that their sheer numbers ensure that predators do not put the overall population at risk.

Within social insect colonies, the workers live briefly as adults but the queens can be very long lived. A queen ant may live for 30 years; a queen termite for up to 50. Mature males are short lived, but queen termites have the same breeding partner (the king) for many years.

In many parts of the world, insects will spend months inactive, as weather conditions are unsuitable (too cold, or too hot and dry). When the insect passes this period in an otherwise active form (larva or adult), this is known as hibernation (in winter) or estivation (in summer) and it can extend the total life span considerably. Adult Brimstone Butterflies (*Gonepteryx rhamni*) hibernate and may thus live for 10 months or so as adults, while most other butterflies in temperate zones overwinter in earlier life-stages and live only a couple of weeks in their adult form.

Some other insects are able to accelerate their life cycle if conditions are favorable. The Vagrant Emperor Dragonfly (*Anax ephippiger*) of Africa is a nomadic species that sometimes completes its breeding cycle much more quickly than normal, allowing it to make use of temporary pools left by heavy rainfall.

Ⓐ A queen termite can live for several decades, during which she will lay millions of eggs.

Ⓥ An adult periodical cicada, which would have spent the first 17 years of its life as a subterranean larva.

## A slow pace of life

In the coldest parts of the world, even summer days may be too cold for much activity. Insects living here tend to have far longer lives than their relatives in warmer climates. Hawker dragonflies (family Aeshnidae) in the far north can take four or five years to reach maturity, while those in warmer countries usually take less than one year. The Arctic Wooly Bear Moth (*Gynaephora groenlandica*), the most northerly-living moth species, takes up to 15 years to complete its life cycle, most of that as a caterpillar.

Ⓐ Because it lives for almost a full year in its adult form, the Brimstone Butterfly has tougher wings than other, shorter-lived species.

# INCOMPLETE METAMORPHOSIS

**Although less dramatic than complete metamorphosis, incomplete metamorphosis still has the power to amaze, as anyone who has witnessed a dragonfly clambering from its larval skin will attest.**

Some insects that undergo incomplete metamorphosis are highly mobile and active from the moment they hatch. Even as young larvae they may be formidable predators, and are certainly better able to escape danger than holometabolous larvae. A bush cricket larva in its first instar may be only a tenth of the size it will eventually attain but it is a near-perfect miniature of an adult, down to its elongated antennae and leaping hind legs. However, it has only very small wing buds and its abdomen is proportionately small as its reproductive organs have yet to fully develop. As it passes through successive molts, its abdomen and wing buds become larger, and (in females) the ovipositor becomes visible in later instars.

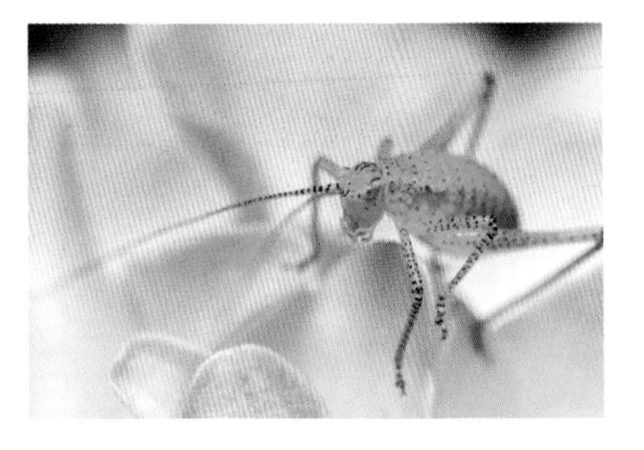

Ⓐ Bush cricket larvae are much like adults but with different body proportions and no wings.

Some hemimetabolous insects hatch as a less developed prolarva stage. The prolarva of a damselfly has no legs and no feeding mouthparts but is capable of limited movement. If it hatches out of water (which can occur if the pond in which its egg was laid has dried up in hot summer weather), it will wriggle its way to water, where it will molt to its long-legged, actively feeding, second instar.

The process of molt is triggered and regulated by the interplay of various hormones. When the juvenile hormone levels drop in the final larval instar, the next molt will be to the fully adult stage. The process of molt is the same as between larval

instars, but in the case of aquatic insects, the larva usually fully exits the water first. Under its larval cuticle, the adult cuticle has developed, complete with whatever modifications are required—gills may no longer be present, and mouthparts may have assumed different proportions.

After it has cracked through its larval cuticle, extracted its antennae and all six legs intact, and climbed free, the newly emerged adult insect will need to transform its tiny, crumpled wings into functional flight-ready structures. It does this by swallowing air, thus increasing the pressure of the hemolymph in its thorax. This forces hemolymph into the wing veins, causing them to expand and stiffen.

## EGG SIZE

It is widely held that hemimetabolous insects lay larger eggs than their holometabolous counterparts, and that their eggs take longer to hatch. This is what we would expect, intuitively, as their hatchlings appear to be so much more developed, and across several insect groups this is indeed what we find.

The lifestyle of larval grasshoppers and crickets remains broadly similar once they reach adulthood —except that they are also ready to breed.

However, there are so many exceptions (due to lifestyle specializations of various kinds) that, on average, there is no difference in egg size—in fact, the largest eggs of all are laid by the holometabolous carpenter bees.

The emergence of a dragonfly. Extricating itself from its larval cuticle requires considerable agility.

# FULL METAMORPHOSIS

**The transformation of plump, crawling caterpillar into colorful, fast-flying butterfly is one of the greatest marvels nature has to offer, and the details of this process are no less marvelous.**

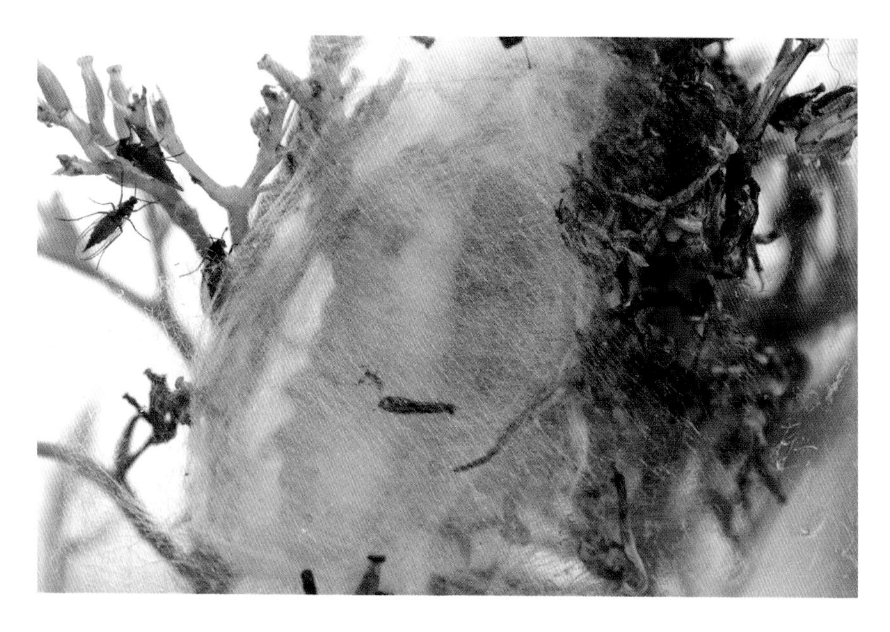

Ⓐ Some caterpillars spin a protective silk cocoon around themselves before pupation.

The pupal stage of development is what separates the holometabolous insects from the more primitive hemimetabolous species. When it enters its pupal stage, the insect molts out of its final larval cuticle to reveal the pupal cuticle beneath. The pupa is typically very different in appearance from the larva. It does not feed and, in most cases, is fixed to the spot and capable only of limited wriggling movement. For this reason, pupae are usually well camouflaged or otherwise hidden.

When an insect larva is ready to pupate, it stops feeding and begins to seek a suitable site for pupation. Among Lepidoptera, for example, hawk moth larvae climb down from their food plant and seek out a patch of soft ground, burying themselves in the earth. Butterfly caterpillars often anchor themselves to a vertical plant stem, or hang by their tail ends from a horizontal twig. Many Lepidoptera larvae use silk produced from their labial glands to stick themselves in position before pupating, and some spin a protective cocoon of silk, within which they pupate. The cuticle of the newly formed pupa soon becomes firm. Those that hang from twigs often resemble dead leaves, while others may be disguised as bird droppings, thorns, buds, or curled-up shards of bark.

Ant, bee, and wasp larvae usually pupate within the nests that their mothers dug or built for them, and within which they spent their larval lives. Social bees, ants, and wasps guard their colony's pupae with vigor, just as they do the eggs and larvae. In honey bee nests, the worker adults seal up the larval cells, once the larvae inside pupate, for extra protection. Some other pupae, though, are far from immobile. Mosquito larvae are aquatic and pupate while still in the water. The pupae, known as "tumblers," float on the surface of the water but can swim and dive if they are threatened. The adult mosquito emerges at the water's surface—it is light and long-legged enough to be supported by the water's surface tension. The larvae of parasitoid insects usually break out of their host's body when mature (causing the death of the already fatally injured host) and pupate on its outer surface, so that the adult insects can fly out into the air when they emerge.

## CHRYSALIDS

Often the words pupa, cocoon, and chrysalis are used interchangeably. However, only pupa is the correct term for the stage between larva and adult in any holometabolous insect. Cocoon refers to the silken case inside which a larva may pupate, while chrysalis is the pupa of a butterfly. The word chrysalis (plural chrysalids) comes from the Greek *chrysos*, gold, as some chrysalids have reflective golden markings. The Cream-spotted Tigerwing (*Tithorea tarricina*) of South America has an entirely shining golden chrysalis, the color produced by the layered structure of its cuticle, which reflects multiple wavelengths of light.

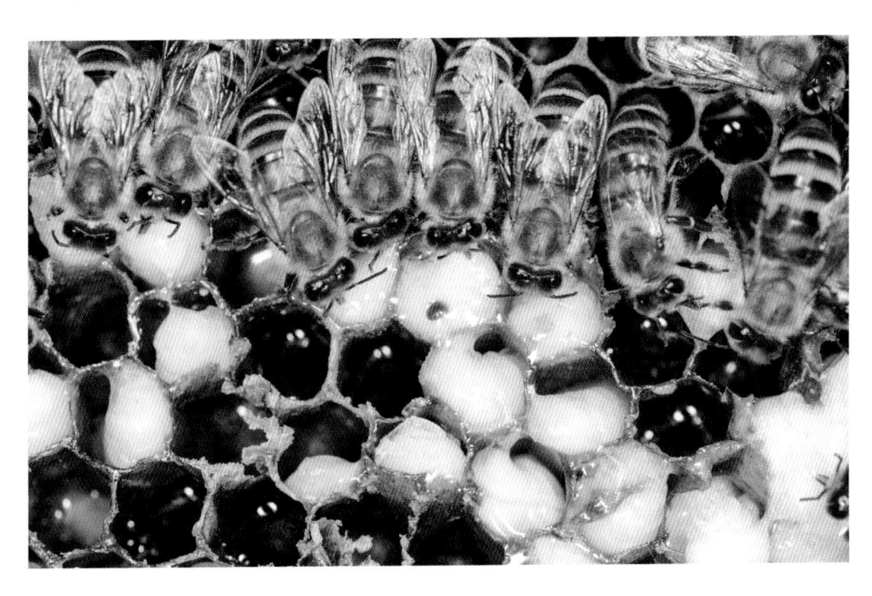

Honey bee larvae pupate inside the cell where they spent their larval lives.

# TRANSFORMATION WITHIN THE PUPA

**Some pupae are quite beautiful, but most are unimpressive, if not downright ugly, to look at. However, that tough and featureless shell hides a minor miracle of engineering occurring within.**

Ⓐ Remodeling a hawk moth caterpillar into its very different adult form requires breakdown and rebuilding of many body structures and tissues.

At first glance, a pupa may bear no obvious resemblance to the larva it was, nor to the insect that it will become. However, it is usually possible to make out the segmented abdomen and the folded wings. Some moth pupae have a long narrow tube on the underside, connected at the head end and partway down the thorax, which will contain the proboscis. The pupae of wasps, bees, ants, and many beetles, are shaped much more like the adult insect, while true fly pupae are variable but often rather featureless egg-shaped blobs. There are two general types of pupae—obtect (appendages are fused to the body wall) and exarate (appendages are free).

Within its pupal cuticle, the insect's larval body parts undergo radical transformation: External wings develop, legs and mouthparts are remodeled, reproductive systems mature. The transformation is far more drastic than in a hemimetabolous insect's change to its adult form. Some or most of the cells that were active in the larva's body will die, while formerly inactive imaginal cells will now begin to divide, multiply, and differentiate into adult tissues. In the case of true flies, whose larvae lack legs and other external features, sac-like structures called imaginal disks within the larva's body will grow into the adult appendages, the central part of the disk becoming the most distal part of the eventual leg, antenna, or mouthpart.

Pupation can be a lengthy process, and the energy needed to kill off old cells and replicate new ones must be fueled entirely by food that the larva consumed before it entered pupation, as a pupa cannot feed. It is also unable to pass waste. Because of this, a newly emerged adult insect may weigh just half as much as it did as a fully grown larva, and one of its first acts is to excrete all of the waste products that have accumulated within its forming body.

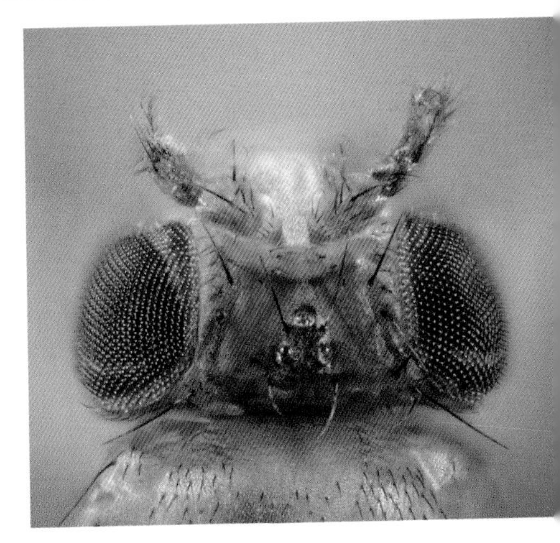

This fruit fly with the genetic mutation "antennapedia" has grown legs where antennae should be.

## HOX GENES

Fruit flies (*Drosophila*) have been used in labs to study genetics for many years. One of the most famous mutated forms of *Drosophila* is a fly with legs growing out of its head instead of antennae. A mutation in a particular gene caused the larval imaginal disk that would normally give rise to an antenna to grow a leg instead. The gene, called "Antennapedia," has been found to belong to a class of genes known as hox genes, which determine which types of appendages will grow out of which segments. Hox genes function as overseers or "managers," determining the general body-plan of the insect by activating or deactivating other genes that build the actual body parts themselves.

Imaginal disk, from which a leg will develop

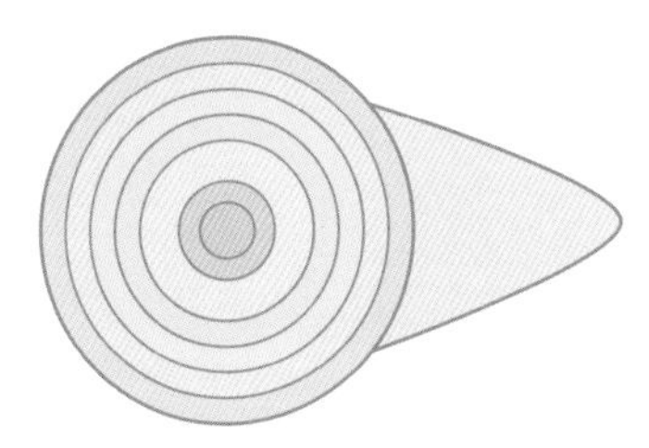

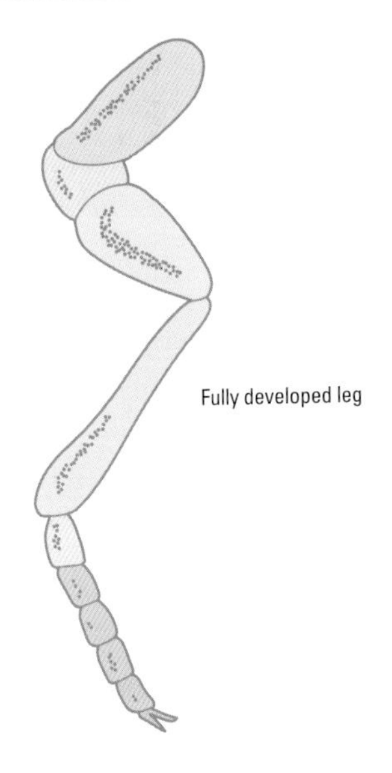

Fully developed leg

The various parts of an insect's leg originate from different "zones" of the imaginal disk.

# EMERGENCE

**Once the adult insect's body is fully formed within its pupa (or within its last larval cuticle, if it is hemimetabolous), it must undertake the last and perhaps most crucial molt of its life.**

Emergence, or eclosure in holometabolous insects, occurs when the adult insect is fully formed, and when it senses that conditions are suitable for emergence. By this stage it is able to see and hear through the shell of its pupal or larval cuticle. Many insects emerge early in the morning, before full daylight, so they have the protection of darkness while at their most vulnerable. Fleas emerge from their pupae when they sense (through vibration) that a living host is nearby.

Some pupae have articulated mandibles, enabling the adult insect to bite its way free. In other cases, in both holometabolous and hemimetabolous insects, the emerging insect begins the process by arching and pushing its thorax upward. The pupal or larval cuticle fractures and the adult insect's thorax appears, soon followed by its head and legs. Once the legs are free it can grip on to a nearby support (often the pupal shell or larval cuticle itself) and finally extract its tiny, crumpled wings and often very distended abdomen.

Once its entire body is free, the insect now spends some time (more than an hour is required for some large species) pumping up its wings and extending its abdomen, taking in air and using the resultant hydraulic pressure to push hemolymph through its system. The wings gradually uncrumple, flatten, and expand as hemolymph fills their veins, and the abdomen elongates, straightens, and narrows.

Although the emerging insect is soft-bodied, unable to move much, and extremely vulnerable to predators, it is also

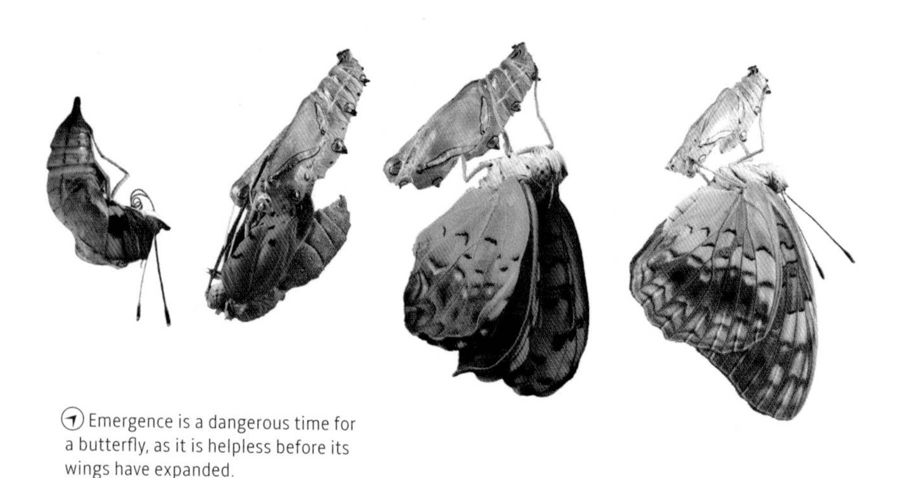

Emergence is a dangerous time for a butterfly, as it is helpless before its wings have expanded.

Ⓐ It benefits mayflies greatly to emerge in large groups, given their brief adult life span and urgent need to find a mate and reproduce.

important that it has air space all around it to allow its wings and body to expand fully before the cuticle becomes stiff. Emergence in deep cover may reduce the risk of predation, but if a twig or other obstacle gets in the way, the insect may end up with a permanently crumpled wing or wings, which would affect its ability to fly, or a bent body that cannot function normally. Other mishaps that can occur during emergence include legs getting stuck and breaking off, and part or all of the pupa or exuvia remaining stuck to a part of the insect's body—the former is a survivable handicap, but the latter may not be.

**MASS EMERGENCE**

Many species of insects show highly synchronous emergence, with large numbers of individuals all taking to the air on the same morning. The advantages of this are clear—predators will be overwhelmed, increasing survival chances of each individual, and the insects will be more likely to find a mate quickly. How, though, do they achieve the synchronicity? Studies on the periodical cicadas, populations of which emerge en masse every 13 or 17 years on the same day, show that, while the cicada larvae only reach maturity at the "right" age, actual emergence is triggered by an outside factor—the temperature of the soil in which they live. Temperature is also the key factor in triggering emergence of winged ants, while studies on mayfly emergence in rivers suggest that a fall in water flow rate may also be involved.

# MATURATION IN ADULTHOOD

**Emergence in its adult form means the insect will no longer molt, but this is not necessarily the end of the process of maturation. Further changes may occur before it is truly a fully fledged adult.**

An insect that is freshly emerged is described as "teneral." In some groups, teneral insects are easy to spot. Dragonflies and damselflies, for example, show muted colors when freshly emerged. Their drab appearance not only provides camouflage from predators but also helps them to avoid competition with and unwanted sexual attention from fully mature adults. Their behavior is also different from mature adults—they immediately move away from the water and spend a day or more hunting in nonaquatic habitats. Only when fully mature, resplendent in their full adult colors, and ready to seek a mate will they return to the waterside. Some immature female damselflies actually have coloration typical of mature males, only developing a female-typical appearance when they are old enough to mate. Copulation is risky for the females and this is only one of several tactics they may use to avoid it (another is to feign sudden death and drop to the ground).

Ⓐ Male *Heliconius hewitsoni* butterflies surround a pupa, competing to mate with the emerging female butterfly.

The activities of an adult worker honey bee change through its life but usually follow the same general pattern. When it first emerges, it cleans out the wax cell in which it developed, and then assists with cleaning other cells nearby. It then helps to feed growing larvae, before moving on to act as a

Ⓥ The subimago mayfly undergoes a final molt to emerge as a fully mature adult.

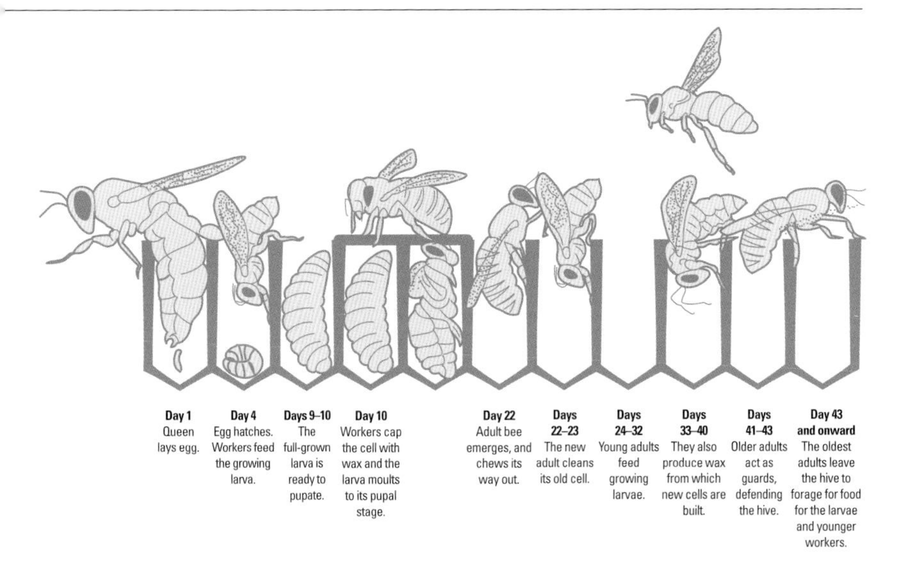


Ⓐ The tasks of a worker honey bee change as it ages—only the oldest bees actually leave the nest to forage for nectar.

guard at the nest entrance. It leaves the nest to begin foraging for nectar and pollen after three weeks of adulthood, and it will then be a forager until the end of its life (probably another two or three weeks later). Any worker bee that is injured during emergence and cannot fly can still be a useful member of the nest workforce.

Female moths and butterflies are sexually mature on emergence, and some even attract a mate before emerging. Aromatic chemicals (pheromones) are released by females still in their pupae, which males can detect. Male *Heliconius* butterflies emerge before females, and home in on a female pupa. Each tries to perch on the pupa and fend off other males, and then tries to mate with the female butterfly as soon as she begins to emerge—this is known as pupal mating. Chemicals that mimic these female pheromones can be synthesized in the laboratory and used to attract male moths of particular species, for purposes of study or pest control.

## WINGED MOLT

Mayflies are unique among insects in that they undergo one final molt in their fully winged form. The so-called subimago or "dun" form, which emerges from the aquatic wingless larva, molts once more soon after it has emerged and flown to a safe place away from the water. It breaks out through the top of the thorax in the usual way, even pulling a fresh set of wings from inside the subimago wing cuticle. Now in its final form, it returns to the water to seek a mate.

10

# BEHAVIOR AND ANATOMY

It is easy to dismiss tiny animals like insects as essentially mindless, but spending some time observing them in the wild will reveal a great breadth of varied and complicated natural behaviors, easily on a par with those shown by so-called "higher" animals, and typically tied into particular anatomical traits.


10.1 · Feeding behavior

10.2 · Breeding behavior

10.3 · Parental care

10.4 · Seasonal behavior

10.5 · Eusocial insects

10.6 · Interspecies interactions


The behavior of social insects, such as the Asiatic Honey Bee (*Apis cerana*), is remarkably complex and elaborate.

# FEEDING BEHAVIOR

**Obtaining and consuming food is a problem that insects solve in many different ways—even among closely related species taking similar foods, there are multiple feeding strategies and behaviors.**

Nectar is a food much used by insects, and is consumed by moths, butterflies, bees, wasps, hoverflies, and others. Most nectar-feeding insects settle on the flower from which they feed, and insert their sucking mouthparts into the flower's nectaries (often loading their own bodies with pollen from the flower's stamens in the process, which they pass on to the female parts of the next flower, thus pollinating it). Some nectar-feeders use a foraging strategy called trap-lining, whereby they patrol the same route repeatedly so that the flower nectaries have time to refill—this requires brains capable of spatial mapping and memory.

Some moths are adapted to take nectar in flight, visiting flowers with pendent (hanging) heads and hovering in front of them as they feed, with their exceptionally long and sturdy proboscises. The feeding manner recalls hummingbirds, and some hawk moths are strikingly similar in appearance to hummingbirds, even down to a flattened fan of hairs on the abdomen tip like a tail, which helps them maintain a steady position as they hover. Bees have chewing as well as licking mouthparts, and some species make use of these to access flower nectaries by biting a hole through the bases of the petals, rather than reaching the nectar through the front of the flower.

Scorpion flies—scavengers that feed on dead insects—are adept at entering spider webs without becoming stuck and feeding on other insects caught in the mesh. They are long-legged and climb deliberately over the strands, holding their wings clear of the web, but will also fight with and sometimes even kill the web's resident spider.

Ⓐ Wasps have biting mouthparts and are capable of dismembering prey as large as themselves.

Ⓐ The crop of a honey bee (sometimes called "honey-stomach") can hold 75mg of nectar (a third of the bee's total weight).

Ⓐ The morphological similarity between a hummingbird and a Hummingbird Hawk Moth (*Macroglossum stellatarum*) is very striking—the moth even has a bristly "tail" to stabilise it in flight.

**HUNTING**

Insects that actively chase down and kill prey are relatively rare, but include the dragonflies and robber flies, both of which are fast flying, with superb eyesight and strong legs to seize and control their prey. Some other predators use camouflage or mimicry to bring their prey to them—these include the orchid mantises, whose flower-like bodies are even more attractive to flower-feeding insects than the actual flowers that they mimic.

Adult workers of the social wasps feed on sweet substances, but they must also collect protein-rich food for their larvae. Because they bring these foods to their nests, they carry out considerable processing to ensure they are only bringing home the choicest parts of the prey. It is not unusual to witness a wasp locked in a struggle with a still-living bee, fly, or moth, the wasp working away with its mandibles to cut off the prey's wings and, often, its head, before flying away with the thorax and abdomen. Wasps will also use their mandibles to cut transportable chunks of meat from a large piece of carrion.

Horseflies of the genus *Haematopota*, the females of which feed on vertebrate blood, are very large for biting flies, but their wings are adapted for silent flight and they have soft rather than scratchy feet, allowing them to approach and settle on their prey undetected. The non-biting males, however, fly with a noisy buzz and have much more conspicuous behavior in general.

# BREEDING BEHAVIOR

**The behaviors involved in courtship, mating, and egg-laying may be very simple, or surprisingly complex. Insects have a very small window of time in which to breed, so devote considerable energy toward getting it right.**

Choosing the right mate can be important, especially for female insects. Males therefore go to considerable lengths to try to improve their chances. One of the methods they use to appeal to a mate (and to warn off rivals) is song. Grasshoppers and crickets make chirping or reeling songs through stridulation—the rubbing together of body parts. In grasshoppers, the sound is produced by rubbing a series of ridges on the hind leg against the forewing, while in crickets the sound is made by the two forewings rubbing together. Cicadas produce their droning whirr of a song using their tymbals, paired membranous structures in the abdomen that vibrate through abdominal muscular action. Different species have their own unique songs, which is a great help in identifying them (especially as they can be remarkably difficult to find).

Patterned wings, especially when only present in the male of the species, are often involved in courtship displays. Male picture-winged flies (family Ulidiidae) raise and flick their boldly patterned wings, and also wave their legs, while approaching a potential mate. This may be enough to convince the female to mate, but she sometimes retrospectively rejects the male by ejecting his sperm from her body (and then eating it). Male demoiselle damselflies (family Calopterygidae) gather at the waterside and perform short showy flights, flicking their dark-banded wings to attract the attention of the plain-winged females.

In most other damselflies there is no courtship ritual—males (which are usually colorful) simply approach the drabber females and try to take hold of them behind the head with their cerci. This position,

Female bush crickets use their prominent blade-like ovipositors to stab a hole into vegetation, into which they lay their eggs.

Ⓐ Male demoiselle damselflies flash their dark-marked wings in courtship displays to attract females.

Ⓛ The loud, tuneless "song" of cicadas is produced through abdominal vibrations.

known as tandem, is the precursor to mating. In some species of damselflies, a proportion of females are andromorphs, meaning they have male-like coloring and patterning and are less likely to be approached by males. Attention from lots of males at the same time can result in injury to the female, and places her at risk of predation, so in years of high population density the andromorphic females often survive in larger numbers than typical females, and still secure enough matings to lay eggs. In other years, the andromorphs may have lower breeding success than typical females. Over time, things even out and both typical and andromorphic females persist in the population.

## REBEL WORKERS

In a honey bee nest, only the queen produces eggs—the workers, though anatomically female, are sterile and exist only to support the nest and the queen's offspring. However, if a nest loses its queen, a certain proportion of workers defy their biological destiny and lay eggs of their own. Because they have not mated, these eggs hatch as male larvae (the same is true when queens lay unfertilized eggs—see page 155). These males leave the nest when mature and could potentially contribute their rebel mothers' genes to a new nest, if they manage to find a virgin queen.

# PARENTAL CARE

**Most female insects do no parenting beyond laying their eggs in a suitable place, and males do even less. However, a few insects do demonstrate lasting and committed care of their young.**

Earwig females are unusually devoted parents. A male and female pair live together in a nest chamber, but after mating, the female chases away her male partner. She lays her eggs in the chamber and remains with them over the week they take to hatch, defending them from potential predators and carefully cleaning them to protect them from fungus. When they hatch, she feeds the larvae on regurgitated food, and they may also eat her if she dies before they leave the nest.

Female *Diploptera punctata* cockroaches also care for and feed their young. Rather than regurgitating food, they secrete a protein-rich, milk-like food for them from specialized abdominal glands known as brood sacs. This food is consumed by the unborn embryos (this species of cockroach is one of the few that is viviparious—gives birth to live young).

Shield bug females also take care of their young—indeed, one European species (*Elasmucha grisea*) is known as the Parent Bug for this reason. A female Parent Bug lays her egg cluster on a birch or alder leaf and shelters it with her body. She places on the eggs the bacteria that the larvae will need to digest plant matter—the bacteria reach the developing larvae through the eggshell. The mother bug stays with the small larvae when they are newly hatched, "herding" them back to the cluster if they try to stray. She only leaves them when they reach their third instar and separate from their family group.

Although solitary bees and wasps do not typically tend their eggs or larvae, they do provide them with a home and a supply of food that will sustain them until pupation age—the nest, which may be a burrow or a mud pot, has its entrance sealed to keep out

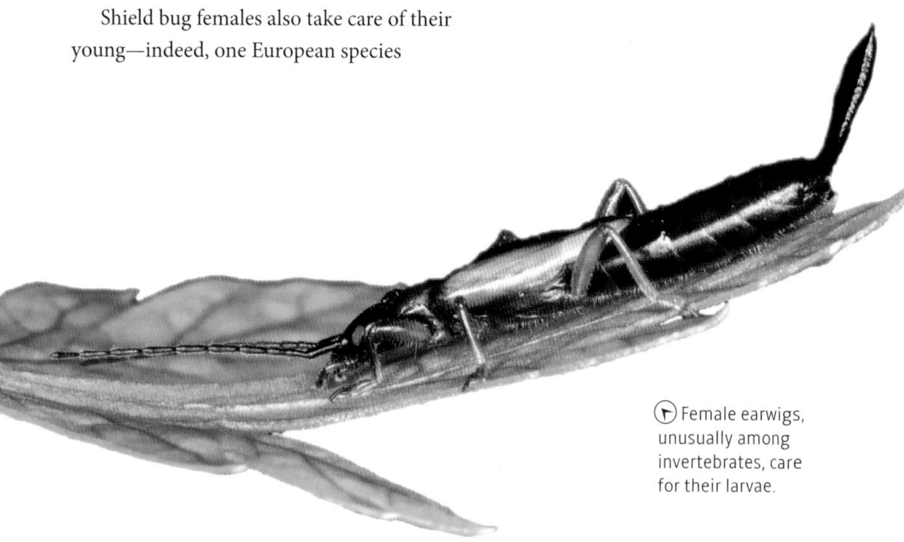

Female earwigs, unusually among invertebrates, care for their larvae.

A male Giant Water Bug, laden with egg masses placed there by females.

A mother Parent Bug stands guard near her cluster of larvae.

predators. It will be provisioned with food—most bee species supply pollen and nectar, while the solitary wasps leave a paralyzed prey item. Some species do remain in attendance after the egg or eggs hatch, providing additional food. This behavior is taken much further among communally living (eusocial) bees and wasps, with adults providing continuous care (see page 168).

## FATHERLY LOVE

The female Giant Water Bug (*Lethocerus deyrolli*) lays her egg mass on her mate's back. He will carry the eggs with him until they hatch, going about his daily foraging activities but also making sure that the eggs are kept wet, and driving away potential predators. A male may mate repeatedly and carry eggs from several females—females prefer to lay their eggs on males who are already carrying some eggs, and thus demonstrating their parenting skills.

# SEASONAL BEHAVIOR

**Seasonal change, whether from summer to fall and winter in temperate areas or the tropical shift from rainy to dry, necessitates different behavioral strategies, as well as anatomical adaptations.**

Most insects, being exothermic, only function well and have full activity within a certain, relatively narrow, temperature range. This means that they rest in a safe place when the temperature is outside that range (often overnight). For those that live in temperate climes, temperature ranges of 104°F (40°C) or more may be expected between summer and winter and an extended period of inactivity may be needed in the coldest (or hottest) periods. Many insects pass the winter months in one of their inactive forms (egg or pupa). Those that overwinter as larvae or adults will enter a period of inactivity (hibernation, or diapause) whereby all metabolic processes are drastically slowed. They may also find or create a shelter (hibernaculum) in which to hibernate, to protect them from the worst of the cold. Some species can tolerate having their body tissues frozen and can overwinter in damp soil, but those that cannot will choose dry hibernacula that are sheltered from the wind.

Ⓐ The beautiful Oleander Hawk Moth (*Daphnis nerrii*) is a rare migrant visitor to Europe from Africa and western Asia.

In hibernation, all developmental processes cease, and metabolic function slows almost to nothing. Hibernating insects consume virtually no oxygen, and bodily fat stores are eked out over the cold season. The physiological "decision" to enter a state of hibernation is determined by genes and triggered by various environmental cues, including hours of daylight, air temperature, and changes in food supply (fewer fresh leaves for herbivores, reduced numbers of

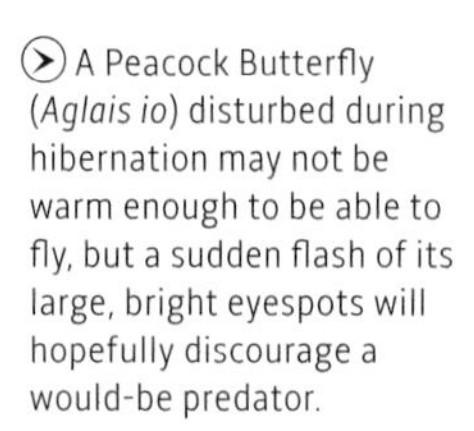

Ⓐ A Peacock Butterfly (*Aglais io*) disturbed during hibernation may not be warm enough to be able to fly, but a sudden flash of its large, bright eyespots will hopefully discourage a would-be predator.

Ⓐ Butterflies that hibernate in their adult form have drab, camouflaged undersides, though the wings' upper surfaces may be colorful.

Ⓐ A dragonfly can keep cool by "obelisking"—this posture reduces the amount of surface area exposed when the sun is directly overhead.

prey for carnivores). When it is readying itself to become active again, the insect can help raise its internal temperature by vibrating its wings, and by basking in sunlight. Large-winged insects change their wing position in response to air temperature, spreading their wings to absorb more heat and closing them to conserve it. Dragonflies adopt a tail-up position called obelisking to avoid overheating on hot days—the posture reduces the surface area exposed to the sun.

Some insects hibernate communally, benefiting from safety in numbers. Ladybugs often form large aggregations inside buildings. They leave chemical compounds as they walk, which guide other ladybugs to the hibernaculum. Moth larvae of the genus *Choristoneura* spin silk shelters in foliage to protect themselves. Migration is another tactic to avoid bad weather, though covering the necessary distance is possible only for strong-flying insects in their adult form.

## TOO HOT

Inactivity in hot, dry conditions is known as estivation. This may occur in the dry season in tropical areas, and in high summer in warm temperate areas. The main hazard facing the insects that estivate is water loss through evaporation. When estivating, insects can reduce water loss in this way by drastically slowing down their respiration rate. However, they can revive to a normal, active state much more quickly than insects that are disturbed from hibernation. The Winter Ant (*Prenolepis imparis*), found across North America, is an example of an insect that is adapted to activity in low temperatures—it estivates in hot summer weather, when most other ants are at their most active.

# INSECT MIGRATION

*When a habitat is hospitable only at certain times of year, the insects that live in it will usually hibernate or estivate through the lean months. Another solution is to migrate.*

Migration is most familiar to us as a bird behavior. Many species breed in temperate regions during the spring and summer and move to or beyond the Equator for the winter months. This behavior is largely driven by insects—most migratory birds are insectivores, and insects are inactive in cold weather, whether they winter as eggs, larvae, pupae, or adults.

Because of their short adult life spans, relatively few insects are capable of undertaking migrations of this distance—breeding is their priority. However, some strong fliers do make long journeys. The most famous of them is the Monarch Butterfly (*Danaus plexippus*), populations of which migrate in the fall from southern Canada and northern and central United States to Florida and Mexico, or down the western side of the Rockies to Southern California, where they form huge gatherings at overwintering sites. In spring, a return journey takes place, but these butterflies are not the same individuals that set out the previous fall—up to four generations are born and die over the course of the journey.

Several insects that mainly occur in Africa, including the Vagrant Emperor Dragonfly (*Anax ephippiger*) and Painted Lady Butterfly (*Vanessa cardui*), migrate north into Europe and western Asia in summer when their populations are unusually high in their normal range. The year 2019 saw millions of Painted Ladies reach the UK, but most years numbers are much lower. The butterflies will breed in their new home, and there is

The mass migration of Monarch Butterflies through the southern USA and Mexico is one of nature's great wonders.

evidence from radar records that the new generation heads back south in fall, traveling at high altitude. Their breeding cycle is rapid and their migratory pattern flexible, allowing them to make the most of local conditions.

The migratory Vagrant Emperor Dragonfly is a rare visitor to the UK but has made the journey from Africa even in the depths of winter.

## Vagrancy

Insects that occur outside their usual range have often been introduced by people, deliberately or accidentally (for example, in imported food or timber). But some insects stray huge distances more or less under their own steam, albeit perhaps with considerable wind assistance. A handful of Monarchs cross the Atlantic to western Europe during the fall most years, and the migratory Green Darner Dragonfly (*Anax junius*) has also made the crossing on occasion.

Most lost migrants will die and leave no mark, but once in a while, vagrancy can lead to colonization. Monarch butterflies are rare in mainland western Europe but are more likely to make it to the Canary Islands. The species become established across the archipelago in the 19th century after several mass arrivals—though its long-term survival was only possible because its larval foodplant (milkweed) exists on the islands.

Painted Ladies migrate from North Africa into northwestern Europe every summer.

# EUSOCIAL INSECTS

**The social organization of bee, ant, wasp, and termite colonies is truly remarkable. Studying how such colonies function reveals how layered and complex insect behavior can be.**

Eusociality is rare in the animal world. It is a system whereby several (perhaps thousands) of individual adult insects cooperate to manage a nesting space, sharing care of young and other tasks. Usually only one or a few of the adults actually reproduce, and the "workers" may belong to distinct "castes," each with its own particular anatomy and social role.

The most famous and well-studied eusocial animal is the honey bee. Within a nest, all eggs are laid by the queen, who is larger than the worker females. There are no different worker castes, but a worker bee's role changes as it ages (see page 155). Once it is foraging outside the nest, it brings back pollen and nectar as food for the larvae. Inside the nest, workers share information

Ⓐ The European or Western Honey Bee (*Apis mellifera*) is the most well-known of the world's 10 or so species of eusocial honeybees, and the only one that has been fully domesticated.

Ⓥ Ants follow the same scent-marked pathways as they head out from their nests to forage.

➤ Termite mounds are imposing structures that offer homes to various other organisms besides termites.

about foraging sites through their "waggle dances"—a figure-eight movement whose orientation and speed corresponds to a location in the outside world. The crop of the honeybee is exceptionally expandable, and when full of nectar can make up a third of the insect's total weight.

Olfactory communication is very important in eusocial insects. A pair of brain structures called the *corpora pedunculata*, or mushroom bodies, are key to learning and memory through scent signals, and are well developed in eusocial insects. Worker ants establish foraging trails, and others follow the chemical cues left behind along these trails. This enables them to make their way directly back to their nest over 330 feet (100m) or more. They will assist fellow workers that are struggling with a heavy burden—in this way, ants can kill prey much larger than themselves and carry it back to the nest.

Termites belong to a different biological grouping from other eusocial insects but have much in common with ants in particular. They build enormous nest mounds from their own droppings and other materials, including clay-rich soil. The mounds are oriented to ensure optimal temperatures inside. Termites feed mainly on dead and decomposing plant material and may encourage and maintain colonies of edible fungi within the nest.

## BREEDING AND SPREADING

An insect colony may persist for many years, though those of social wasps and bumblebees only survive for one season. Among ants, new reproductive females with wings leave the nest and mate with a male or males from another nest. After this, they find a site to establish a new colony and shed their wings. The fertilized eggs they lay become new female worker ants. In social wasps, new queens mate in fall and then hibernate. The following spring, they find a nest site and build the foundations of a new nest from chewed wood fibers. The queen lays a few eggs and feeds the larvae when they hatch. Once they mature into adult workers, they take over the work of adding to the nest and caring for the queen's subsequent offspring.

# MUTUALISM AND COMMENSALITY

*There are numerous examples of mutually beneficial relationships between two species in nature. However, in many relationships only one of the two benefits, neither helping nor hurting the other.*

Ⓐ Ants tend their aphid "herds" with great care, as the aphids provide them with a generous supply of honeydew.

Mutualistic or symbiotic relationships among insects are exemplified by the link between ants and aphids. Aphids are true bugs, which feed on plant sap by puncturing the plant with their piercing mouthparts. They excrete excess fluid, and this "honeydew" is loaded with sugar, which provides food for many other species—for example, several species of tree-dwelling butterflies feed mainly on aphid honeydew in their adult stages. Ants also value aphid honeydew and often corral and guard "herds" of aphids, protecting them from predators and harvesting their honeydew.

When one species relies on another for its survival, but does so without affecting the "host," the relationship is called commensalism. This is distinct from mutualism where both species benefit, and also from parasitism, where one species is harmful to the other. Pure commensality is rare in nature. However, large ant nests often host a number of other species that live apparently unnoticed in the nest, feeding on detritus and not bothering the host insects in any obvious way. The large "gallery" tunnel complexes used by certain wood-boring beetles also provide home, shelter, and food for various other insects and invertebrates, which exploit the wood that decays around the edges of the tunnels.

There are many mutualistic relationships

where one partner appears to get a lot more from the arrangement than the other. One example is seen in burying beetles, and the tiny mites that they often carry on their bodies. These beetles lay their eggs on corpses of vertebrate animals, as do the mites. The mites use the beetles as transportation, and in return they help reduce competition for the beetles at carcasses, by killing off fly eggs and larvae. However, the mites do also eat the burying beetles' eggs and larvae.

Ⓐ Termite activities enrich the soil, allowing different plant communities to grow than in surrounding areas.

## Housing estates

The huge, robust mound nests made by certain species of termites provide homes for many animals. The nests are made from compressed soil particles, hardened by the baking sun. Termite mounds in the African savanna can reach tremendous size—more than 100 feet (30m) across. These huge mounds often support more trees than the surrounding areas, as their soil is more fertile, providing patches of rich habitat that can sustain many other animals.

Some termites make their nests underground, with a network of tunnels linking various nesting chambers to each other and to the surface. Their nests and tunnels may be shared by mites and springtails.

Ⓥ Burying beetles provide a taxi service for mites—both lay their eggs on carrion.

# INTERSPECIES INTERACTIONS

**Insects prey on many other animals—including other insects. They also parasitize them, live in their homes, and even, in a few instances, have mutually beneficial relationships.**

Beyond the simple links between predator and prey, it is among the social insects that we see some of the more complex interspecies relationships. Ants are well known for herding and guarding colonies of aphids, in order to drink the honeydew that the aphids secrete. Other insects also benefit from the protection and care of ants, but in the case of the Large Blue Butterfly (*Phengaris arion*) the ants are duped into giving far more. When a Large Blue caterpillar reaches its fourth instar, it drops down from its food plant and is quickly found by ants of the species *Myrmica sabuleti*. Fooled by pheromones and sounds produced by the caterpillar, the ants take it into their nest and treat it as a larval queen ant, feeding it in preference to the colony's own larvae (even sometimes feeding it on their own larvae) and protecting it from danger, until it pupates and leaves the nest as an adult butterfly.

Phoresy is the transport of one species by another, and is usually benign in nature. Water mites hitch rides between water bodies by riding on damselflies, and scavenging mites travel on the backs of burying beetles. Parasitoid wasps of the genus *Trichogramma* use phoresy in a more sinister manner. They can sense anti-aphrodisiac pheromones released by female butterflies of their host species. The butterflies release these pheromones after they have mated, to discourage further male attention. But to the wasps, the pheromones indicate that the butterfly is likely to lay eggs soon. On detecting the pheromone, the wasp will climb on to the butterfly's body, so as to be ready to lay its own eggs on the butterfly's eggs when they are laid. Mantisflies carry out a similar trick with spiders (see page 139), though in this case it is the larvae that hitch a ride, rather than the adult females.

Through pollination, insects enable plants to reproduce sexually and to access a wide gene pool.

Ants assiduously guard their "herd" of aphids and move them to good feeding areas.

## QUEENS AND SLAVES

Several species of bees are brood parasites, laying their eggs in other bees' nests. Female cuckoo bumblebees (*Psithyrus* spp.) actually enter and live in the nests of their host bumblebee species, laying many eggs and sometimes even attacking and killing the resident queen. The resident workers take care of the cuckoo bee's eggs as if they were their own. The slave-maker ants take over other nests' workforces in an even more dramatic manner. Workers of these ant species seek out nests of other ants, and abduct pupae from them, sometimes several thousand, to bring to their own nests and raise as co-opted workers.

The slavemaker ants have specialist workers which are tasked with finding other ant nests to attack. Once one of these "scouts" has found a suitable nest, it goes back to its own nest, leaving a scent trail, which others will follow back to the targeted nest to carry out the raid, usually meeting very little resistance. The ants that emerge from the stolen pupae imprint on their new colony completely, and may even end up carrying out raids on their original nests.

The Large Blue Butterfly lays its eggs on thyme plants (*Thymus* spp.), but the caterpillars switch to a more carnivorous diet in their final instar.

# HOST–PARASITE RELATIONSHIPS

*Even the most healthy-looking human being may still be host to a variety of tiny parasitic animals, both internal and external. Parasitism is a fact of life for virtually all living things.*

A parasite is defined as an organism that lives on or in another organism, feeding on its body tissues in some way, and usually causing some harm (though this is not necessarily serious or life-threatening). The parasite depends on its host for survival, at least during one of its life stages, if not its entire life. Parasites that live on the outside of the body are known as ectoparasites, while those that live inside are endoparasites.

Some insects live as parasites, while others are hosts to them, and some species may be both host and parasite at the same time. A few insects are parasitic on humans, among them the fleas. Most species of flea have a preferred host, but will take blood from other vertebrates as well—the so-called Human Flea (*Pulex irritans*) has dozens of recorded hosts other than humans, and the Cat Flea (*Ctenocephalides felis*) will happily bite people as well as cats. The Chigoe Flea (*Tunga penetrans*), found in tropical Central and South America and also (as a non-native species) southern Africa, affects humans and other species and actually burrows into the skin, creating painful lesions. The Head Louse (*Pediculus humanus*) and the Pubic Louse (*Pthirus pubis*), however, are obligate parasites of humans. Unlike fleas, which only habitually live on their host's bodies in their adult form, the lice are full-time residents, laying their eggs on head hair and pubic hair respectively and feeding on blood.

Insects themselves are infected by a range of endoparasites, especially protozoa (single-celled organisms), which the insects usually take into their bodies while feeding. The protozoa may cause considerable

Ⓐ Larvae of the fly *Dermatobia hominis* are parasites of humans and other mammals.

Ⓐ Most blowflies (family Calliphoridae) lay their eggs on carrion or dung, but some lay eggs on wounds on the bodies of living vertebrate animals.

damage by feeding on internal body tissues, and a heavy infestation can kill the host. Some parasitic protozoa are used as natural control agents to reduce populations of harmful insects. Other kinds of protozoa, including the species that causes malaria, need to be passed on to another organism to complete their life cycle.

## The Human Botfly

This species of fly, *Dermatobia hominis*, found in Mexico and South and Central America, is a particularly alarming parasite of humans and other mammals. The female botfly captures a female mosquito and affixes an egg to the mosquito's mouthparts, where it hatches. In this way, when the mosquito bites human skin, the botfly larva is transferred to the wound and burrows deep inside. The larva feeds on its host's tissues and grows to its full size over the following eight weeks, before (assuming the host has not managed to remove it) falling out to pupate in soil. Removing a living larva is very difficult, as it resists being pulled out and if its body breaks in the attempt the wound will probably become infected. If the entry wound is covered with petroleum jelly the larva will suffocate after several hours, and can then be extracted with a pair of tweezers.

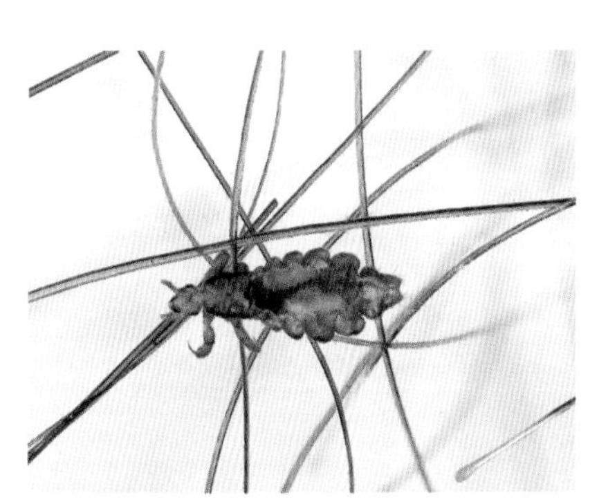

Ⓐ The Head Louse is a tenacious parasite of humans and is found almost everywhere on Earth.

Ⓐ Fleas can spread serious diseases, including typhus and bubonic plague.

# PARASITOIDS

*Insects are not always very invested in care of their young. The parasitoids are an exception to this, but their idea of good parenting is many people's idea of the scariest horror story.*

Ⓐ Large White butterflies are the host for the parasitoid wasp *Cotesia glomerata*.

Parasitoidism is a particularly gruesome form of parasitism, in which the host is eventually killed by its parasite, but not before the parasite has used the host's living body as food and shelter for an entire stage of its life. The majority of known insect parasitoids are in the order Hymenoptera—in particular the ichneumon, braconid, and chalcid wasps.

Female parasitoids may use many different hosts, but many specialize in one or a few host species and attack it at a particular life stage. For example, the female braconid wasp *Cotesia glomerata* of Europe and Asia injects her eggs into the bodies of caterpillars from the genus *Pieris*, particularly the Large White (*P. brassicae*). The parasitized caterpillar appears to behave normally but inside it the eggs hatch and the wasp larvae feed on its body tissues. After a couple of weeks, the larvae emerge from the caterpillar's body, killing it, and form their pupae around it—although sometimes they themselves are attacked before they can pupate by another specialist parasitoid—the ichneumon *Lysibia nana*. In some cases, parasitoid attacks can have dramatic impacts on their hosts' population, and the populations of parasitoid and host tend to rise and fall year on year in a regular cycle.

Some parasitoid wasps, such as the potter or mason wasps (family Vespidae), construct a nest and place one or more of their hosts inside it, stinging them first to

Ⓐ All insect life stages are vulnerable to attack from parasitoids, Here, two adult parasitoid wasps emerge from parasitized stink bug eggs.

Ⓑ Parasitoid larvae chew their way out of their host's body before they pupate in a cluster around the remains.

paralyze them and prevent their escape. The wasp lays an egg in or on each host's body, and the larva consumes it gradually.

## Internal battles

Hosts attempt to thwart parasitoids in various ways. Some try to escape their attackers, and many have an immune response to the injected eggs. The parasitoids are very persistent and in some cases lay so many eggs in the host that the immune system is overwhelmed. The act of egg-laying sometimes also introduces a virus to the host's body, which compromises its immune system.

In a few cases, the parasitoids can actually modify the host's behavior. For example, host caterpillars parasitized by endoparasitoid wasps of the genus *Glyptapanteles* will try to defend the wasp pupae that form around its dying body.

# CELLS AND BIOCHEMISTRY

Cells are the building blocks of all life and in this respect insects are no different from other animals. Each individual cell is a self-contained little biological machine, with its own particular role to play and tasks to carry out in order to keep the entire insect healthy and functional. Cellular processes involve interactions between individual molecules—in the realm of biochemistry.


11.1 • Structure of a typical cell

11.2 • Cell organelles

11.3 • Cell replication

11.4 • Immunology

11.5 • Specialized cell types

11.6 • Insects in cellular research


A butterfly alighting on our hand is a treat, but this behavior usually means it is interested in consuming the salts and other minerals contained in our sweat, to meet certain cellular requirements.

# STRUCTURE OF A TYPICAL CELL

**Cells are the self-contained, microscopic structures from which insects' and other animals' bodies are built. They are differentiated into many types, each with its own "job" to carry out.**

We think of insects as small organisms, but compared to single-celled organisms such as amoebae, insects are very large and very complex. The cells that make up animal bodies are, on average, the same size, regardless of the actual size of the whole animal. Each gram of animal tissue contains about 1 billion cells, so a honey bee weighing one tenth of a gram has some 100 million cells in its body. The heaviest of all living insects, the Giant Weta (*Deinacrida heteracantha*) of New Zealand, can weigh 2½ ounces (75g), which means its body contains about 75 billion cells.

Although a cell is self-contained and to some extent self-supporting, it is part of a community of similar cells, which work together, forming the distinct organs and systems within an insect's body. Most cells have the means to replicate themselves, allowing for growth in the larval stages, and some level of tissue repair and regrowth in all stages.

Cells vary in structure according to their function—nearly all of them are quite specialized in one way or another. However, most have at least some commonalities. The contents of a cell are known as its cytoplasm. This is contained within a flexible cell membrane. The membrane is made of layers of phospholipid molecules. It is semipermeable—certain molecules can pass through it, but most cannot. As they work, cells consume oxygen and release carbon dioxide, and these gases can diffuse through the membrane. Larger molecules tend to pass through at specific points where protein-based structures (channels or

↓ The cell membrane is formed by a double layer of phospholipid molecules, plus other molecules involved in detecting and transporting nutrients and other substances into and out of the cell.

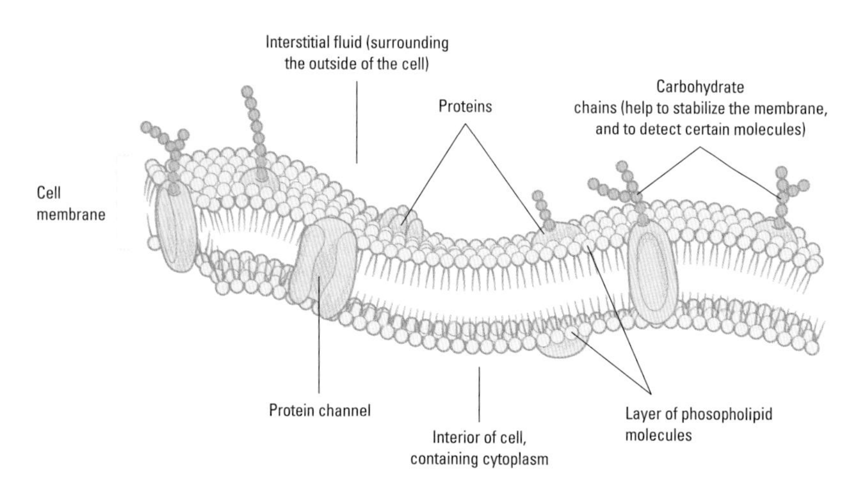

transporters) are embedded in the membrane. Molecules may also be absorbed into the cell by being engulfed by an inward fold of the membrane.

As well as its fluid component (the cytosol), the cytoplasm includes a number of smaller structures called organelles—they are the working parts of the cell, like organs within an organism. The number and type of organelles varies according to the type of cell. Their roles are described on pages 182–183.

**RECEPTORS**

The cell membrane often also has molecules on its outer surface that bind chemically to hormones, neurotransmitters and other molecules. They do not enter the cell, but cause a change in the cell's state. For example, a neurotransmitter binding to a neuron (nerve cell) causes electrically charged ions to enter the cell, and this electrical charge then passes along the cell.

Ⓐ The acute and fast-acting sensory systems of damselflies are formed from various different highly specialised cell types.

Ⓥ Insects' body cells are fluid-filled, and the cell membrane's control over fluid balance enables insects to survive even in arid environments.

# CELL ORGANELLES

**Under the microscope, several distinct structures are visible inside a typical animal cell. Some are large and prominent, while others can only be made out under the highest magnification, but all have important functions.**

The most prominent structure inside a cell is its nucleus, which looks like a dark circular blob. This organelle is the cell's control center, determining and overseeing its other activities. The cell's genetic material—its paired chromosomes—reside within the nucleus. Chromosomes are strands of DNA that hold all the instructions telling the cell which proteins to build. When a cell divides, the process begins with the chromosomes in the nucleus being duplicated, a process that also relies on another type of organelle—the centriole. A distinct darker area within the nucleus, the nucleolus, is the building site for ribosomes, another type of organelle, which is directly involved in actually building proteins.

The membrane of the nucleus is combined with another membranous structure called endoplasmic reticulum (ER). This membrane comes in two forms—rough ER, which has ribosomes bound to it and is involved with protein-building, and smooth ER, which lacks ribosomes and builds fat molecules from free fatty acids. There are also ribosomes free in the cytoplasm. The Golgi apparatus is another membranous structure, which is involved in fine modifications to proteins built by the ribosomes.

Mitochondria are relatively large, oval organelles found within the cytoplasm. Their role is the creation of the energy-providing molecule ATP, through the metabolism of oxygen and glucose. They contain their own DNA (known as mitochondrial DNA or mtDNA), and can replicate themselves using this DNA.

Growing larvae need to replicate their body cells rapidly.

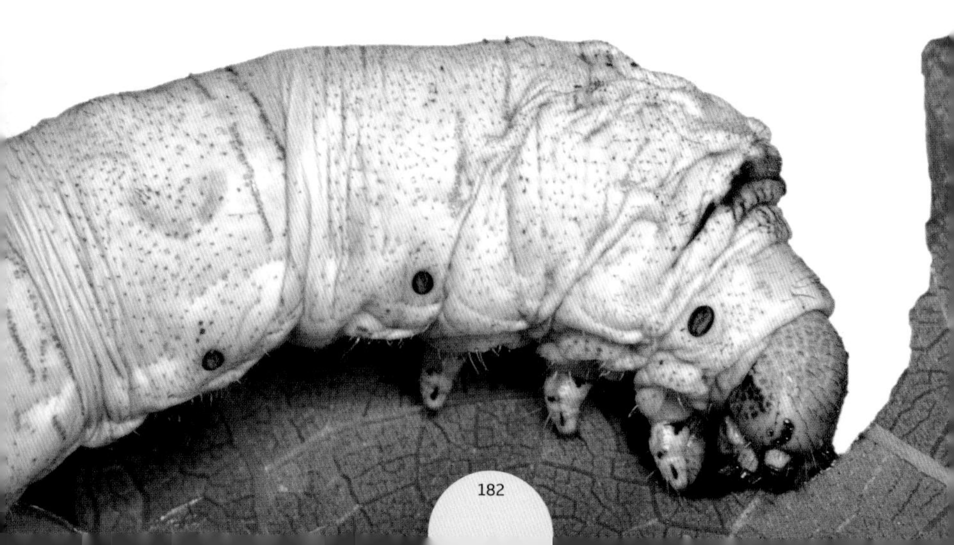

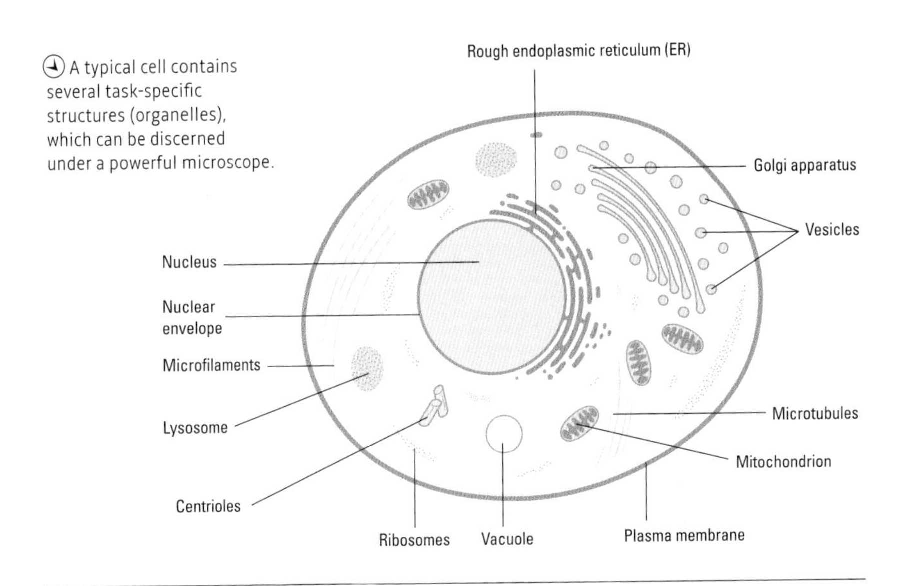


A typical cell contains several task-specific structures (organelles), which can be discerned under a powerful microscope.

The cytoplasm may also contain fat stores, in membrane-bound vesicles. Water-soluble molecules are stored in fluid-filled pockets called vacuoles. The lysosome is a particular kind of vesicle (sac) that contains enzymes. These enzymes help break down waste products and obsolete remnants of spent organelles, converting them into molecules small enough to pass out of the cell through its membrane.

## NUMBERS OF ORGANELLES

Cells vary in terms of how many organelles of each type they contain. Cells with a high rate of energy consumption contain more mitochondria than average. Sperm cells, for example, contain a cluster of mitochondria near the tail end, to power its swimming movement. Some of the longest muscle cells have multiple nuclei, as do some types of large cells carried in the hemolymph and involved with the immune response.

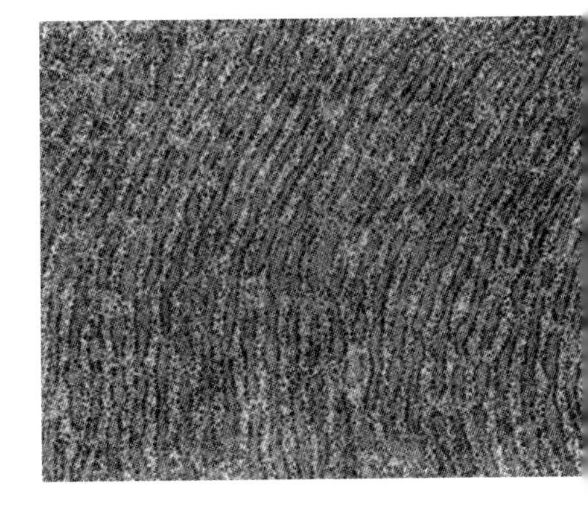

Under the scanning electron microscope, the ribosomes attached to the folds of rough endoplasmic reticulum are visible.

# CELL REPLICATION

**For single-celled organisms, splitting into two is the means of reproduction. Higher organisms reproduce sexually, but this process, as well as ordinary bodily growth, also begins with cell division.**

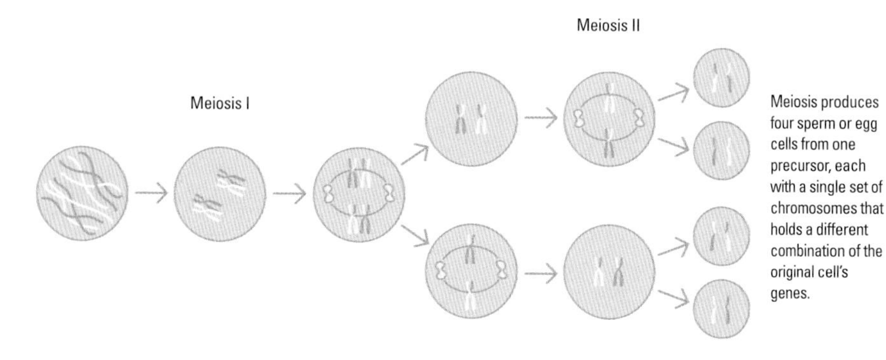


The chromosomes have been duplicated, so the cell nucleus contains a double set. They now begin to separate, and crossing over begins.

Crossing over continues as the chromosomes line up in their pairs.

Homologous chromosomes line up in pairs.

The chromosomes split and move to opposite sides of the nucleus. The nucleus—and then the cell itself—divides into two.

Within each of the two new cells, the chromosomes line up again, but this time without being duplicated first.

Each of the new cells splits again, but this time the nucleus contains only one set of chromosomes rather than two (it is haploid rather than diploid).

Ⓐ Meiosis is the process by which sperm and egg cells form, each carrying 50 percent of the parental genes but in different combinations.

All insects begin their lives as a single fertilized cell, so the number of cell divisions that they go through during their lives is astronomical. The rate of division in the earliest life of an embryo is extremely rapid—in fruit fly embryos, one cell has become 6,000 within about three hours of fertilization.

The first stages of normal cell division involve the entire cell becoming larger, its organelles moving into particular parts of the cell (and sometimes duplicating), and the creation of a copy of all of the chromosome pairs inside its nucleus. It also makes a duplicate of its centrosome, the organelle that will separate the two sets of chromosomes. This process is known as interphase.

The second phase, mitosis, involves the two sets of chromosomes shortening and becoming aligned within the nucleus, and at the same time the two centrosomes form a spindle-shaped structure around the nucleus. This spindle pulls apart the two sets of chromosomes. Once they are separated, the chromosomes reassume their normal elongated, disorganized form, and the nuclear membrane forms separate seals around each set, forming two nuclei in place of one. As the two nuclei stretch and part, the entire cell membrane is also stretched and begins to pinch inward in the middle, eventually forming two separate cells.

As cells divide, they may also begin to differentiate, from generalized stem cells into those adapted for a specific function. Differentiation is gradual, stem cells becoming fully specialized cells via one or more intermediate cell types, or precursor cells. The first precursors to neurons, for example, are neuroblasts, which then become ganglion mother cells and then ganglion cells. These ganglion cells differentiate into either neurons or glial cells (another cell type found in the nervous system).

## MEIOSIS

The process of cell division is different when the cells being formed are the gametes or sex cells—the ova and the sperm. Each ovum and each sperm should contain only one chromosome from each pair, and when fertilization occurs a full chromosome set is formed by the combination of each gamete's half-set. Cell division from the gametes' precursor cells therefore involves an additional stage whereby each precursor cell divides twice, producing four daughter cells (ova or sperm), each with a half-set of chromosomes. This process is called meiosis. Another important part of meiosis is the "crossing over" that occurs in the earliest stages, when the tangle of chromosomes in the nucleus of a precursor are lined up in their matched pairs. During this sorting, both chromosomes in a pair will break at certain points (chiasmata) and the broken-off sections will join the other member of the pair at the same point. Although pairs of chromosomes are made of the same genes in the same sequence, they may have different "versions"—alleles—of these genes. Crossing over results in different combinations of the alleles in each sperm and egg cell.

The developing embryo of a fruit fly, showing body segmentation. This stage is about 8 hours after fertilization.

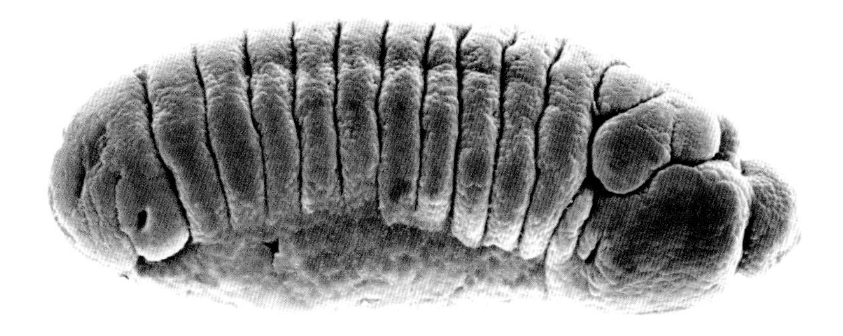

# IMMUNOLOGY

**Insects protect themselves from disease through behavior, and by the integrity of their body structure. If infectious agents like bacteria do find their way into the body, an immune system response occurs.**

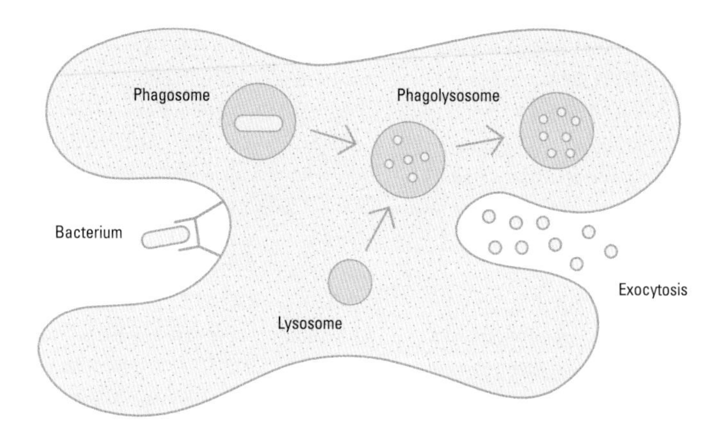


Ⓐ A phagocytic cell engulfs a bacterium, forming a phagosome. An enzyme-filled lysosome combines with the phagosome, forming a phagolysosome in which the bacterium is broken down. The products of its breakdown are then released (exocytosis).

Any adult or larval insect's body can potentially be invaded by disease-causing viruses, bacteria, fungi, protozoa, and, in some cases, parasites and the eggs of parasitoid insects. The first line of defense, the cuticle, has its vulnerabilities, such as the spiracles that let air into the body, and any injury to the cuticle creates a significant new vulnerability. Pathogens could also be eaten accidentally, or even introduced during copulation.

Insect hemolymph contains specialized cells that detect and respond to foreign bodies. They deal with small objects such as bacterial cells by engulfing them (phagocytosis), and then breaking them down. They form aggregations around larger objects, such as eggs injected into the insect's body by parasitoid wasps, in an attempt to fully encapsulate it. These cells are called hemocytes, and come in three main types, of which the most numerous are phagocytic plasmatocytes. As well as engulfing pathogens, these cells release signaling molecules, which alert other hemocytes to the danger. The other two types are crystal cells, which release molecules that attack pathogens, and lamellocytes, which are involved in the encapsulating process.

As well as an army of hemocytes, the insect produces pathogen-attacking proteins in its fat body, which are released into the hemolymph. These protein molecules attack pathogens such as fungi and certain types of bacteria. They are also involved in forming clots in the hemolymph at the sites of wounds.

Insects do not have specialized cells in their hemolymph that provide an acquired immune response. In vertebrates, this function is carried out by B-lymphocytes and T-lymphocytes, and these cells help ensure that some infectious diseases, if successfully fought off, will never take hold a second time. However, insects do show increased resistance to a pathogen that they have encountered before. Studies on honey bees show that the enhanced response can even be passed on from a queen to her offspring. Biologists have not yet established how this response is achieved.

## DISEASE IN INSECTS

We tend to think of insects more as carriers of disease than sufferers, but insects are susceptible to many kinds of diseases and, in common with other animals, their immune response may not be up to the task. Biologists studying insect pathology have developed strains of bacteria, viruses, and fungi that can be used to kill off outbreaks of commercially damaging insects—for example, conservationists have used the Japanese fungus *Entomophaga maimaiga* in North America to control the non-native Gypsy Moth, whose larvae can be very damaging to native tree species. Gypsy Moths have also been successfully controlled in some areas through use of a species-specific virus, which kills the caterpillars. Viral particles are then shed from the dead caterpillars onto nearby foliage.

This fly has died from a fungal infection. The fungal mycelia erupting from between the abdominal segments will release airborne spores that infect new hosts.

# SPECIALIZED CELL TYPES

**All the cells in an insect's body have been derived from generalized stem cells and are adapted to meet a particular function. Some adaptations are particularly extreme.**

Probably the most distinctive type of cell that insects' (and other sexually reproducing organisms') bodies produce is the sperm cell, with its long, fast-moving tail, or flagellum. This projection enables the sperm to swim, through an arrangement of sliding microtubules inside the flagellum, which create the rippling bends that propel it along. Studies on mosquito sperm have shown that the motion of the flagellum is accelerated in the presence of certain chemical signals—effectively, the sperm cell has a sense of smell, thanks to chemical receptor molecules.

The neuron or nerve cell is another distinctive, elongated cell type, with an extended axon along which nerve impulses travel to the next neuron (see page 58). However, they are only part of the story when it comes to the nervous system. There are also very large numbers of glial cells in both the central and the peripheral nervous system. They come in several types, including star-shaped cells that take up unused neurotransmitters around synapses, and sheathing cells that provide protection for the delicate axons.

The fat body contains cells called adipocytes, used for storage of fats. These round cells hold their stores in the form of droplets and can expand to a great size when necessary—for example, in larvae during their final instars, to store fuel that will be consumed during the process of metamorphosis.

Muscle cells (myocytes) have the ability to shorten (contract) and lengthen again when they relax. They contain fibers of two different types of protein (actin and myosin), which break their chemical bonds to slide

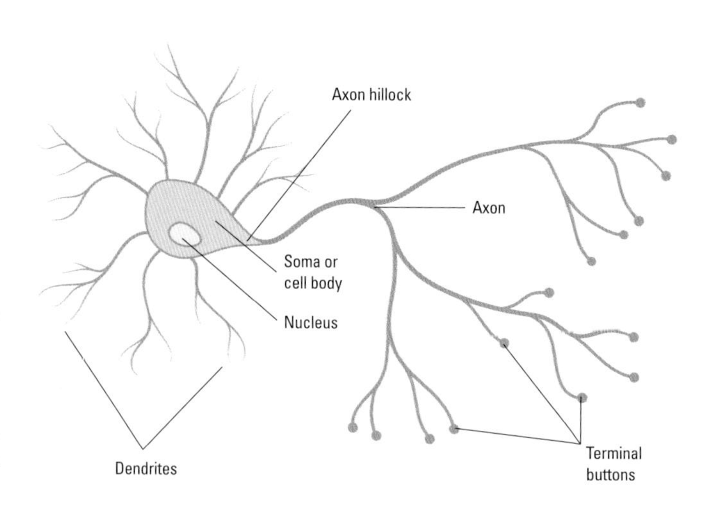


The neuron or nerve cell has a fine, branching structure to transmit electrical impulses from dendrites to the axon's terminal buttons. From there, the impulse passes to the next neuron and thus around the body.

Chafer beetle larvae build large fat-stores over their growth period, which can last up to five years.

over each other when the muscle contracts. Muscle myocytes have an elongated structure and a very high supply of mitochondria to fuel their energetic activity.

## SINGLE-CELL GLANDS

The epidermis of an insect contains large, specialized secretory cells that function as exocrine glands, meaning they secrete their product outward, rather than inside the body. Exocrine glands produce various different chemical compounds, including the pheromones with which insects may attract a mate, and unpleasant-smelling or bad-tasting substances intended to discourage predators. Although insects may have many exocrine glands with this defensive function, it is typically only those closest to the site of stimulation that will emit the secretion.

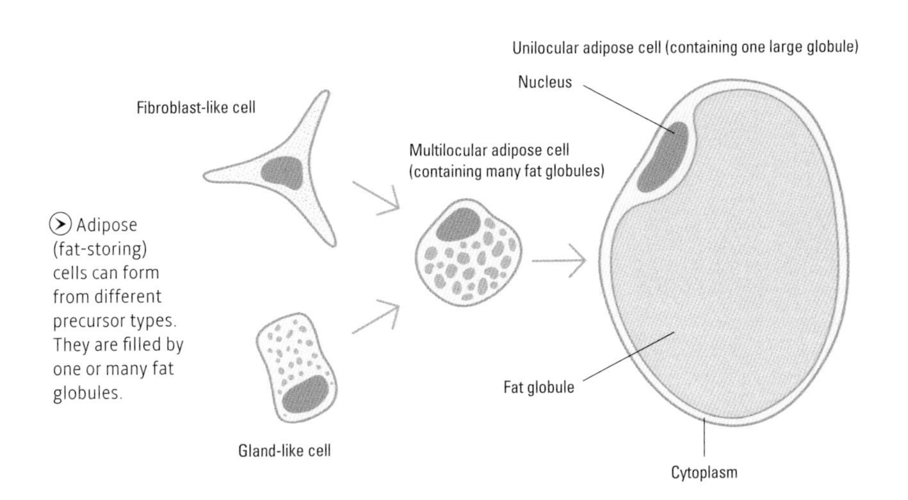


Adipose (fat-storing) cells can form from different precursor types. They are filled by one or many fat globules.

# INSECTS IN CELLULAR RESEARCH

**Because of their small size and how easy it is to keep and breed them, insects are popular study animals in laboratories. We owe much of our understanding of cell biology to insect subjects.**

Ⓐ An insect's head and compound eye. Scanning electron microscopes can produce amazingly detailed close-up images of insect body parts, and even of individual cell organelles.

Studying cells of any kind became possible with the invention of the microscope. The first cells were observed under a light microscope in the 17th century. This device uses several lenses to magnify the image, and material to be viewed is placed in a thin layer on a glass slide, lit up from below. The use of staining chemicals reveals structures within the cell more clearly (for example, mitochondria are not visible at all without some kind of staining).

The electron microscope is a much more powerful device for visualizing tiny details at high magnification and resolution. First invented in the early 20th century and revised and refined many times since, this device uses a beam of electrons rather than visible light to generate its images. Through electron microscopy of cells, the detailed structure of organelles can be studied. Although cell research has concentrated primarily on plant and vertebrate cells, since the 1970s new methods have been developed to prepare insect cells for microscopy and our understanding of insect cell biology has made great advances.

Lineages of identical, cloned cells in the laboratory are used frequently for studying cellular biochemistry. Using identical cells ensures that experiments can be repeated

Ⓐ The larval form of the Fall Armyworm Moth—the species from which the widely used cell lineage Sf9 was derived. The cells' many uses include study on why cells 'self-destruct' at a predetermined age.

reliably, without any outside variables being introduced as a result of genetic differences between the cells used. Several lineages of cells derived from insects are used in biochemistry laboratories for work such as using cells to build particular types of proteins and studying their response to different viruses.

The lineage Sf9, derived from ovary tissue taken from the Fall Armyworm Moth (*Spodoptera frugiperda*), is used in labs for a wide variety of purposes, including development of influenza vaccines, a study of how cell function changes in low-gravity conditions, and investigations into the genes that control apoptosis (programmed cell death), which is involved in aging processes.

**HIDDEN HISTORY**

One of the most interesting revelations to come from investigations into insect cell biology concerns the traits that are still shared between them and us. For example, the molecular processes that guide the development of egg and sperm cells have been found to be largely the same in insects and mammals. This indicates that this process came into being more than 500 million years ago.

# DIVERSITY AND CONSERVATION

The importance of insects in the world's ecosystems cannot be overstated—we and all other vertebrates are heavily dependent on them. Millions of different insect species exist, each with its own role in the environment, but today insects are dying out at an alarming rate, and active conservation has never been more vital.


12.1 • Types of insects

12.2 • Insect communities in different habitats

12.3 • Record-breakers

12.4 • Threats facing insects

12.5 • Extinction

12.6 • Insect conservation


➤ The Colorado Potato Beetle, native to parts of North America but accidentally introduced to many other areas, is the world's most significant insect pest of potato crops.

# TYPES OF INSECTS

**All of the insects on Earth belong to one of 30 or so main groupings, or orders. Some of these orders are very familiar to us, and contain many thousands of species.**

There are four insect orders that dominate the rest in terms of species diversity, with at least 100,000 species each. They are the beetles (Coleoptera), the true or two-winged flies (Diptera), the wasps, bees, ants, and sawflies (Hymenoptera), and the butterflies and moths (Lepidoptera). All of these four are holometabolous (going through complete metamorphosis) and they can be found in all continents except Antarctica.

Other sizable insect orders include the true bugs (Hemiptera) with about 75,000 described species worldwide. This order is the most diverse of all of the hemimetabolous (partially metamorphosing) insects, and also the most diverse, with great variation in body shape and lifestyle. Behind the true bugs come the grasshoppers, crickets, katydids, and relatives (Orthoptera) with nearly 20,000 species, and the caddisflies (Trichoptera), with some 14,500 known species. Other well-known insect orders include the cockroaches and termites (Blattodea—about 7,400 species), the dragonflies and damselflies (Odonata—about 5,500 species), the lacewings and their relatives (Neuroptera—some 4,000 species), the stoneflies (Plecoptera—about 3,500 species), the stick insects and their relatives (Phasmatodea—about 2,500 species), the mantises (Mantodea—about 2,300 species), and the earwigs (Dermaptera—about 1,000 species).

Examples of all of these orders can usually be found quite easily in most parts of the world, though less so the further you depart from large landmasses in tropical areas. Each group has its own distinctive traits and, despite the vast number of species, it is usually relatively straightforward to work out which is the right order for any unknown insect you encounter.

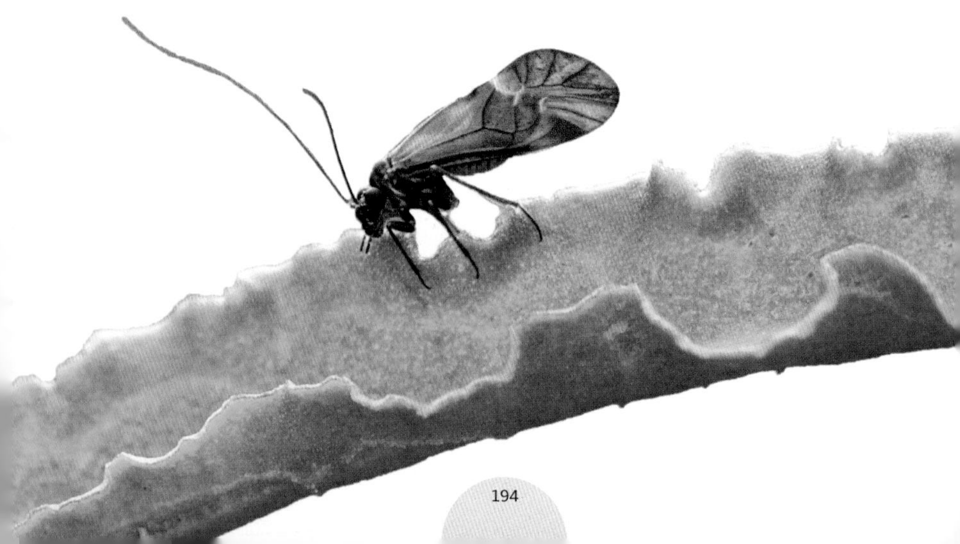

A barklouse, one of the members of the little-studied order Psocoptera.

Ⓐ The highly diverse order Hymenoptera includes some impressive predators, such as the elegant *Ammophila* hunting wasps.

Ⓐ The snakeflies, belonging to the order Raphidioptera, are distinctive, large-winged insects with elongated thoraxes.

## OBSCURE ORDERS

The remaining insect orders are much less known by casual and amateur nature-watchers, because they are fewer in number and more difficult to observe. Among them are the web spinners (Embioptera), about 200 species of mostly tropical, four-winged insects that live in the soil or other substrate, inside webs and tunnels spun from silk, which they produce from glands in their front legs. The extremely tiny insects that form the order Strepsiptera resemble true flies in that they have only one pair of functional wings, the other being reduced to small clubs (halteres). All 600 or so species are parasites of other insects. The 1,000 or so booklice and barklice (Psocoptera) are small, primitive insects with grinding jaws. Booklice are wingless and mainly live inside human dwellings, where they can damage stored foods and other organic material (including the pages of books). The biting and sucking lice (order Phthiraptera) are wingless, strong-legged parasites of birds and mammals—the biting lice chew on skin, hair, or feathers, while the sucking lice suck blood or other bodily fluids.

Ⓡ The order Neuroptera includes several very striking families, including the predatory, rather dragonfly-like owlflies (family Ascalaphidae).

# DOMESTICATED INSECTS

*Relatively few wild animal species have been kept in captivity long enough and in large enough numbers to be classed as fully domesticated, but among them are some very economically important insects.*

Ⓐ A single hive of Honey Bees can produce 60 pounds (27kg) of honey over a season.

The best-known domesticated insect is the European or Western Honey Bee (*Apis mellifera*). It is one of several social bee species that produces honey in its large nests, and humans have a long history of exploiting it and its relatives in the wild. The invention of beehives and an understanding of how to control and manage the bee's natural behavior led to its domestication, with the first evidence of deliberate beekeeping coming from North Africa (where the species naturally originated) some 9,000 years ago. Today, honey bees are used not only to produce honey and other products like beeswax and royal jelly, but as crop pollinators—hives are moved around as needed to ensure the bees pollinate the target plants. The total commercial value of honey bees worldwide is around $200 billion per year.

The Silk Moth (*Bombyx mori*) is the domesticated form of the wild species *B. mandarina*. It is a large moth, originating from Japan, China, and Korea, and is kept for the cocoons its larvae spin around their pupae. The threads they secrete from their mouth glands are used to weave silk, which developed into a lucrative trade. Each cocoon may yield a mile of thread, but because the moth damages the threads when it emerges, commercial silk-farming involves killing the pupae before this occurs, except for a small

Ⓥ The domesticated Silk Moth has long since lost the ability to fly.

Ⓥ The Silk Moth's wild relatives include the spectacular Giant Peacock Moth (*Saturnia pyri*).

number from which the adult moths are used for breeding.

Like many domesticated animals, Silk Moths are quite different from their wild ancestors. Their wings have reduced in size and have lost their natural camouflaging pigments. They will also feed on other foods besides mulberry leaves (the sole food plant of the wild species). Deliberate selective breeding produced these traits, which make the moths easier to keep in captivity. Research is underway to modify the Silk Moth genome in such a way that, instead of silk, their caterpillars secrete different proteins, with other commercial or medical uses.

Ⓐ Much of what we know today about animal genetics comes from research on the Common Fruit Fly (*Drosophila melanogaster*).

## Model organisms

Another domesticated insect is the Common Fruit Fly (*Drosophila melanogaster*), famous for its use in genetic research. The fly is ideal for this purpose as it is very easy to breed in the lab and has a relatively simple genome. Mutated forms have been developed with undersized or twisted wings, and different eye and body colors, and each mutation's position in the genome has been documented.

The Black Soldier Fly (*Hermetia illucens*) is beginning to be used commercially as a food source for pet cats and dogs. Its larvae are protein-rich and may be kept easily at high densities in small spaces. They feed on decaying matter, so can be sustained on waste food of all kinds.

# INSECT COMMUNITIES IN DIFFERENT HABITATS

**Wherever a variety of plants grow, there will be plenty of insects, but it is the nature of the plant community that will dictate which kinds of insects are most numerous.**

Different ways to classify habitat types, or biomes, have been used by different groups over the last few decades. Today, the WWF, or World Wide Fund for Nature, recognizes 14 types of terrestrial (land) biomes, and 12 freshwater biomes. The criteria for definition for the land biomes are, broadly speaking, types of plant cover (forest, woodland, grassland, shrubland, or desert), humidity, altitude, and latitude (tropical, subtropical, temperate, or polar), while the freshwater biomes also take into account the water flow (river or lake). To these types could also be added a range of anthropogenic biomes—habitats that have been extensively modified by humans, primarily urban areas, tree plantations, and croplands.

Natural woodland and forest support the greatest range of insect species. Leaf litter provides foraging and hiding places for ground beetles and ants, herbaceous understory plants support plant-sucking true bugs, and their flowers sustain butterflies, bees, and hoverflies. Tree trunks are full of fissures in which insects can hibernate, decaying wood is food for beetle larvae, and the foliage of different tree species supports a great range of insects, including aphids, moth caterpillars, and bush crickets. An abundance of plant-eaters attracts predators. A tree plantation may look just as lush as a natural forest but supports far fewer species of animals, as it typically contains only one species of tree, and usually has no understory and no decaying wood to speak of.

Grassland is less species-rich than forest but scrubby grassland in particular can include large numbers of flowering plants, and thus is very attractive to far-ranging nectar-feeders such as larger butterflies, moths, and bees. Where grassland is close to water, it is also often home to maturing newly emerged dragonflies and damselflies.

Freshwater habitats are home to certain insect groups that lay their eggs in water—among them dragonflies and damselflies, caddisflies, stoneflies, alderflies, and mayflies. Different species have different preferences—some favor cold upland streams, others slow-flowing, well-vegetated lowland rivers. Some water-dependent species are adapted to breed rapidly and make use of temporary, seasonal pools and puddles.

### INSECTS IN GARDENS

Gardens vary individually, but in general are similar to light woodlands in their vegetation community, with trees, shrubs, and clearings. However, many gardens are kept much "tidier" than natural habitats, with little ground cover, and their flora includes many non-native species, which will support few, if any, native herbivorous insects. The widespread use of chemical pesticides can dramatically reduce insect populations.

Many insects spend their larval stages underwater, while others feed on plants that grow in freshwater habitats. Wetland species include the Common Picturewing dragonfly.

Flower-rich meadows support many nectar-feeding insects, which in Eurasia and North Africa may include the Marmalade Hoverfly.

The Southeast Asian planthopper *Pyrops candelaria* is one of many striking insects found in tropical forest habitats.

# INSECTS IN EXTREME CONDITIONS

*Habitats with warm temperatures and an abundance of food sources support the widest variety of insect species, but a few are adapted to live in much more challenging conditions.*

Optimal habitats may meet all of an insect's needs, but they have one significant disadvantage—competition. The consequences of competition are that species tend to become more highly specialized, and that makes them vulnerable. A moth that only lays its eggs on one species of plant may be exquisitely adapted to exploit that host plant, but it will not survive if the host plant dies out.

The Arctic Wooly Bear Moth (*Gynaephora groenlandica*) occurs in Greenland and Arctic Canada. The caterpillar, like those of other tiger moth caterpillars, has long hair on its body. Completely inactive for most of the year, it can endure temperatures as low as -94°F (-70°C) within its silk-lined shelter or hibernaculum, thanks to chemicals in its cells that protect them from damage when they freeze. On the warmest days, it basks in sunshine to absorb heat, which speeds up its metabolism. Once mature, it pupates, emerges, and breeds over just four weeks, in the height of the Arctic summer. Larvae of the Antarctic Midge (*Belgica antarctica*) survive freezing by becoming dehydrated, preventing the formation of damaging ice crystals.

Living on the bodies of other animals presents a particular set of challenges, especially when the host animal is a small,

Louse flies spend much of their lives clinging to the feathers of fast-flying birds such as swifts.

fast-flying bird that is also a long-distance migrant. The Swift Lousefly (*Crataerina pallida*) lives on the blood of swifts, and its larvae overwinter in swift nests after the birds set off on migration. Although the louse is a large insect, its very flattened body means that it can slide under the bird's feathers and thus remain attached while the bird is airborne (on flights that may last days and involve speeds of up to 68 miles per hour/110kmh).

## Desert baskers

Several African beetles of the family Tenebrionidae are adapted to survive in the Namib Desert. This hot, dry environment encourages rapid dehydration, but these beetles have physical adaptations to minimize water loss. A couple of species have also evolved ingenious ways of obtaining water from the early morning mists that sweep the dunes. The most famous is the Head-stander Beetle (*Onymacris unguicularis*). These beetles climb to the summit of a dune before dusk and stand on their front legs, hoisting their rear ends into the air. Fog condenses on their backs and runs down to their mouthparts. They are capable of consuming large volumes of water—up to 40 percent of their own body-weight.

Ⓥ Some desert-dwelling tenebrionid beetles have bumps on their elytra, where fog will condense.

Ⓐ A Head-Stander Beetle in the Namib Desert: This species performs early morning handstands on its front legs to encourage fog to condense on its back.

# THE ECOLOGICAL ROLES OCCUPIED BY INSECTS

*As the most abundant group of land animals on Earth, in terms of diversity of species, insects play many vital roles within ecosystems in all kinds of habitats.*

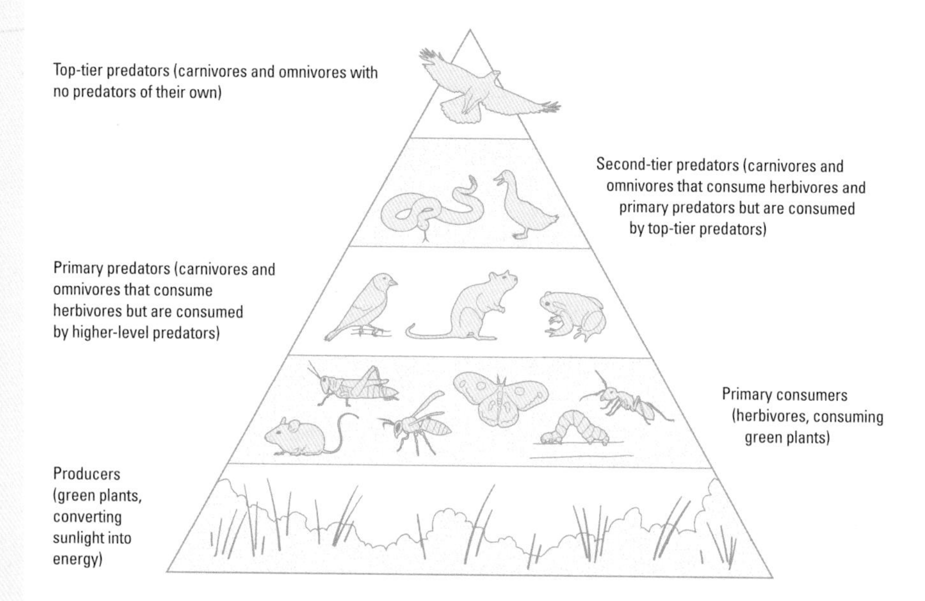


Ⓐ This pyramid diagram, though simplified, indicates how energy from sunlight passes from plant to herbivore to carnivore. With each stage, some of the energy is lost—therefore the biomass of lower-tier organisms in any environment will always be much higher than that of higher-tier organisms.

All life needs energy, and animals obtain theirs through eating other animals and plants. This energy's original source is our Sun. Through photosynthesis, green plants and bacteria-like microorganisms called cyanobacteria can use sunlight to drive a chemical reaction in their cells that turns water vapor and carbon dioxide into glucose—a simple sugar that is the basic energy supply for living things. Photosynthesis enabled plant-eating animals to evolve, and therefore animal-eating animals too.

Ecosystems are built upon "producers," which photosynthesize. On top of that layer are the "consumers," which obtain their energy by eating the producers. There is then an opportunity for another tier of consumers—the predators, which eat the first-tier consumers. The producers must greatly outnumber the consumers for the system to remain stable, which is why ecosystems on tiny isolated islands can be so vulnerable to destructive one-off events. But in a stable ecosystem, there are opportunities for species to become more specialized, and for biodiversity to increase as evolution progresses. The result is not a neatly tiered cake but a complex web of life, incorporating not just herbivores and carnivores but also omnivores, parasites, scavengers, and more.

## Niches

The ecological position an animal occupies—where it lives and what it consumes—is termed its niche. For example, a diving beetle is a predator of small animals in freshwater habitats. A dragonfly larva has a similar niche, but the adult dragonfly occupies a different niche, being an aerial predator of other flying insects. Niches can overlap, but where the extent of overlap is considerable, competition between the niche-holders exerts evolutionary pressure toward more extreme specializations.

Insects occupy all kinds of ecological niches. Some species are particularly influential and important within their ecosystems—without these so-called "keystone species," the ecosystems would be drastically different. Insect keystone species include termites, which create islands of soil fertile enough to support miniature forests in savannas, and aphids, which excrete honeydew on tree foliage that sustains myriad other insects, especially when nectar is in short supply. Insects such as blowflies and burying beetles are also vital disposers of carrion and animal droppings.

Ⓥ A rare example of an insect hunting a vertebrate—a dragonfly larva captures a tadpole.

# INSECTS AS BIOCONTROL AGENTS

*When one species causes problems for us, we have sometimes attempted to solve it by introducing another species to control the first. This method can be effective, but can also go horribly wrong.*

Humankind's early attempts to practice biocontrol (using one living thing to reduce the population of another) have resulted in some significant ecological problems. In New Zealand, people introduced stoats to prey on the large numbers of previously introduced rabbits, reasoning that, because stoats are effective predators of rabbits in Europe, the same would apply in New Zealand. The result, though, was that the stoats preferred to hunt native New Zealand birds, which, having never encountered stoats in their evolutionary history, were entirely ill-equipped to cope with the new danger. Now, both stoats and rabbits are problematic invasive species in New Zealand.

The important lesson to learn from this and similar case histories is that biocontrol is highly risky unless the controlling agent is adapted to attack only one target species. This is where insects are perceived as reliable agents—specifically, the parasitoid wasps and flies that lay their eggs on or in the bodies of just one host species.

Today, several parasitoids have been used successfully to control particular problematic insects. In French Polynesia, the non-native Glassy-Winged Sharpshooter (*Homalodisca vitripennis*), a large plant-eating leafhopper, has been reduced in number by 95 percent by the wasp *Gonatocerus ashmeadi*. The parasitoid wasp *Encarsia formosa* was used as a controlling agent for the Glasshouse Whitefly (*Trialeurodes vaporariorum*) in commercial greenhouses from the 1920s, and although its use was reduced dramatically when pesticides became widely available from the mid-20th century, it is now enjoying a considerable revival as many growers seek to reduce pesticide use. Biocontrol agents like these can be used by certified organic farmers and growers, while chemical pesticides cannot.

## Natural biocontrol

Gardeners who want to reduce or end their reliance on pesticides use can help achieve this goal by encouraging pest insects' natural, native wild predators. We tend to think of birds rather than insects in this context, but many predatory insects can be highly effective at keeping down numbers of plant-eating pests. Ladybugs and lacewings both hunt aphids, both as larvae and adults. Providing hibernation sites for them in the garden can help ensure that they stay around. Hoverflies should also be encouraged in gardens for their pest-control abilities—the larvae of several species prey on aphids and, as adults, hoverflies are important pollinators.

A whitefly pupa: These insects are serious greenhouse pests—biological control against them has been used for more than 100 years.

The Glassy-winged Sharpshooter is a plant-eating true bug which can cause significant crop damage—biological pest control is used against it in French Polynesia.

Hoverflies are pollinators in their adult form, but as larvae some species prey on leaf-eating insects, making them valuable to gardeners.

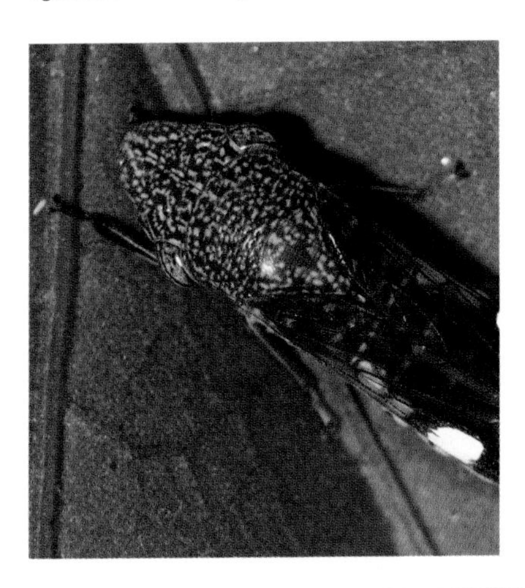

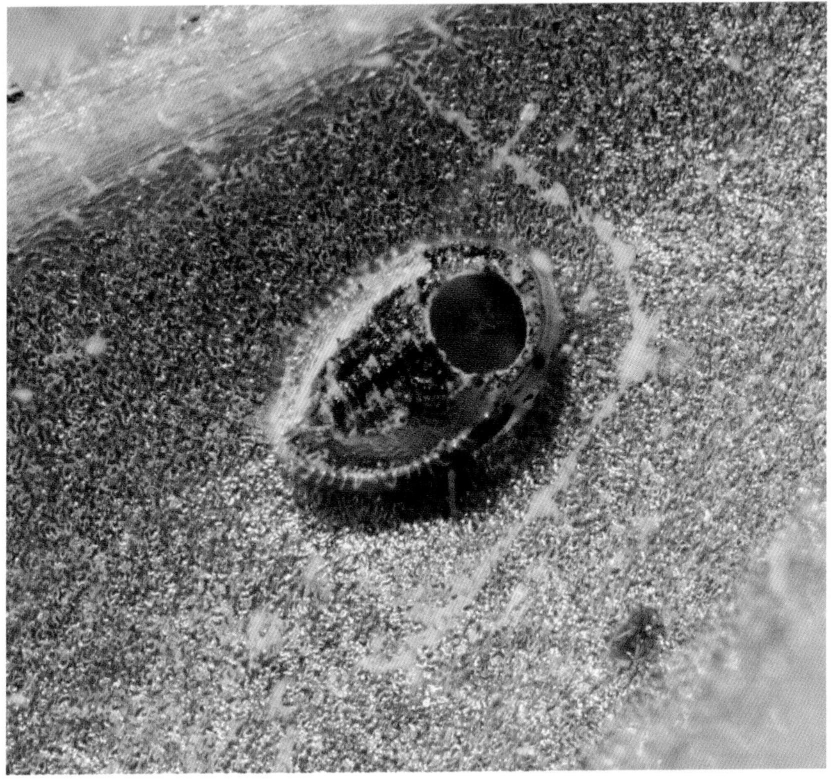

# RECORD-BREAKERS

**Although every insect species has its own impressive qualities, a few have earned themselves a place in the record books for traits or abilities that set them apart from all the rest.**

Insects stand out among other animals in many ways. They are the most abundant, so arguably the most successful of animal groups. They are the only flying invertebrates and the only animal group to show advanced eusociality.

In terms of record-breaking individual species, the heaviest insect is the Little Barrier Island Giant Weta (*Deinacrida heteracantha*), which can weigh 2½ ounces (75g). The longest-bodied is the extremely spindly stick insect *Phryganistria chinensis*, which can reach more than 24 inches (62cm) in length. The Hercules Beetle (*Dynastes hercules*) is the longest-bodied beetle at up to 6⅔ inches (17.5cm), with more than half of that length made up of its enormously long thoracic "horn." The smallest insects are male wasps of the family Mymaridae, which can be just 0.14mm long.

The White Witch Moth (*Thysania agrippina)* has the broadest wingspan of any insect—up to 12 inches (30cm)—but its wings are relatively narrow, and in terms of wing surface area it is beaten by other moths, including the Atlas Moth (*Attacus atlas*).

As we have seen elsewhere, the fastest-flying insects are the dragonflies, with large hawk moths hot on their heels, while the fastest runner is the Australian tiger beetle *Cicindela hudsoni*, and whirligig beetles (family Gyrinidae) are the fastest swimmers. The Common Froghopper (*Philaenus spumarius*) holds the high jump record, leaping upward more than 27 inches (70cm). In terms of distances traveled, some individuals of the Globe Skimmer Dragonfly (*Pantala flavescens*) are known to travel as far as 3,730 miles (6,000km) on their migratory journeys.

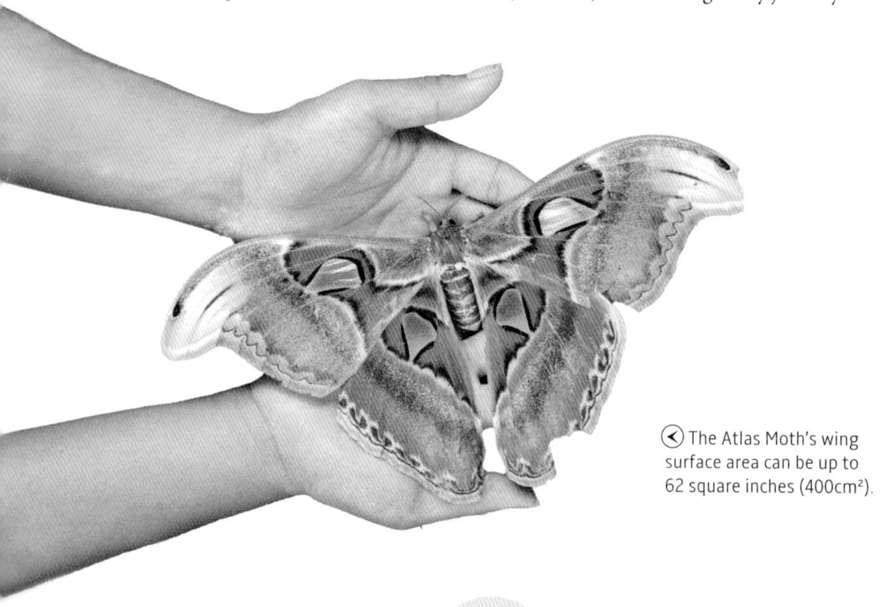

The Atlas Moth's wing surface area can be up to 62 square inches ($400cm^2$).

(A) A White Witch Moth is inconspicuous when resting, despite its tremendous 12-inch (30cm) wingspan.

(<) The Hercules Beetle can measure $6^{2}/_{5}$ inches (17.5cm) from the tip of its horn to the tip of its abdomen.

The loudest insect is the African cicada *Brevisana brevis*, males of which chirp at 110 decibels. The insect with the brightest bioluminescence is the Headlight Elator Beetle (*Pyrophorus noctilucus*)—like other bioluminescent animals, it produces its light (from a pair of "headlights" on its thorax) through the action of the enzyme luciferase on the light-emitting pigment luciferin.

## EXCEPTIONAL ANTS

There are somewhere between 10 trillion and 10,000 trillion ants alive on Earth today, making them the most abundant insect group in terms of numbers of individuals. Their total biomass is about the same as the Earth's human biomass. Some ant species form nests of extraordinary size. A colony of Japanese Red Wood Ants (*Formica japonica*) on the island of Hokkaido, Japan, was formed from 45 individual but interconnected nests, and is estimated to hold about 1.1 million queens and 306 million workers. A colony of non-native Argentine Ants (*Linepithema humile*) in Melbourne, Australia, is about 60 miles (100km) across.

# THREATS FACING INSECTS

**Insects, like all organisms, face a range of threats to their survival. Their populations are so fundamental to ecosystems that their loss often impacts negatively on many other species.**

Earth is a dynamic system, and over its long history many natural events have occurred that have devastated animal populations, on local and global scales. Over the last century, though, most of the serious threats facing insects have come about as a result of human activity.

The greatest single threat is habitat loss and damage. Wild habitats are under constant and increasing pressure from an encroaching human population. They may be directly replaced wholesale, by farmland, plantations, or urban development, or may suffer indirectly, through desertification, pollution, or fragmentation. For an insect population, even a small road built through the area they occupy may be an insurmountable barrier, and when populations become broken up they are more susceptible to the problems of inbreeding (reduced genetic diversity, making genetic problems more likely to occur).

Ⓐ The Yellow Crazy Ant has had a serious impact on the biodiversity of several islands where it has been accidentally introduced.

Some insect species can do tremendous damage to our crops, and this problem is often tackled with pesticides. However, traditional pesticides such as DDT kill all insects, and their widespread use across western Europe and North America in the 1960s and 1970s resulted in devastating crashes, not just in insect populations.

Our unfortunate habit of introducing non-native species to countries far from their origin has also had a severe impact on many insect species. This problem is worst on islands where a restricted fauna has evolved, often in the absence of any higher predators. Adding a generalist predator (one that evolved within a much more diverse and mixed fauna) to such an environment often spells disaster for the native species. For example, in New Zealand (a country that, historically, had no native predatory mammals) several of the 70 weta species are threatened by introduced rats, stoats, cats, and hedgehogs. Weta are large, slow-moving relatives of the grasshoppers—they evolved alongside insectivorous birds, but are ill-equipped to defend themselves against these mammalian predators.

Collecting insect specimens for museums is not generally too harmful if carried out responsibly, but species that are already very rare may be threatened with total extinction by unscrupulous collectors, eager to secure a specimen before it is too late.

## INSECTS AS THREATS TO OTHER SPECIES

Invasive, non-native species threaten wildlife and ecosystems in many parts of the world, and several damaging invasive species are insects. One of the most notorious is the Yellow Crazy Ant (*Anoplolepis gracilipes*), probably native to East Africa but now present on many island groups. It is highly predatory, aggressive, and forms vast super-colonies. Its impact on the Christmas Island fauna has been devastating.

Ⓥ The beauty of New Zealand's landscape belies the extent of the ecological damage brought about by an array of non-native species.

# EXTINCTION

**The vast majority of all species that have ever evolved have become extinct, and this fate awaits the rest, even ourselves. However, today the extinction rate is far higher than is "natural."**

When the last individual of a species dies, that species has become extinct and cannot ever return (at least, not until cloning technology is improved). If the species survives only in captivity it is classed as extinct in the wild, but if the remaining individuals (whether in the wild or in captivity) cannot generate a viable breeding population (if numbers are too low, for example, or if all remaining individuals are of the same sex), that species is considered to be functionally extinct.

Of course, many (perhaps most) extinctions go unobserved by human eyes—and there is no doubt that among those were countless species that were never even "discovered," especially when it comes to insects as so many are small, difficult to observe, and hard to identify.

Confirming extinction in the wild is always somewhat uncertain, and there are many examples of rediscovery of presumed-extinct species. But, in general, if repeated field surveys of all known sites for an endangered species turn up no sightings over several seasons, then that species is likely to have become extinct. Recently extinct insects include the Xerces Blue (*Glaucopsyche xerces*), a North American butterfly last seen in 1943, a victim of urban development of San Francisco in its sand dune habitat. This species gave its name to the conservation

The fossil record shows us that the general extinction rate has fluctuated considerably over the last 500 million years, but has shown five conspicuous spikes, indicating rapid mass extinction events.

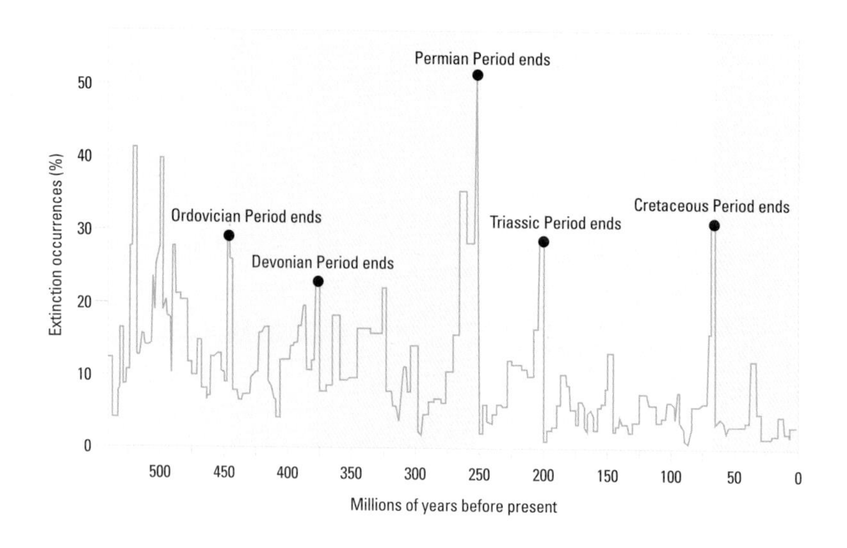

Ⓐ The only way to see the lovely Xerces Blue butterfly today is in a museum cabinet.

group the Xerces Society, established in 1971 and committed to protecting populations of invertebrates considered essential to biodiversity and ecosystem health.

Other recently extinct insects include *Megalagrion jugorum*, a spectacularly long-bodied damselfly from the Hawaiian island of Maui, the once extraordinarily abundant Rocky Mountain Locust (*Melanoplus spretus*) of North America, whose swarms could include more than 10 trillion individuals, and the Saint Helena Earwig (*Labidura loveridgei*), an enormous earwig endemic to Saint Helena island in the South Atlantic, which was wiped out in the 1960s by introduced rodents and predatory invertebrates.

**RETURN FROM OBLIVION**

Cloning offers a real chance of restoring an extinct species to the world. The technique would involve extracting a cell nucleus (complete with a full set of DNA) taken from preserved specimens of the extinct species, and placing it inside an egg cell of a very similar host species. The result would be a new living animal that was a genetic "twin" of the specimen from which the cell nucleus was taken. The first successful cloning of an insect took place in 2004, using fruit flies (*Drosophila*) but the technique has not yet been used to try to recreate an extinct species.

Ⓥ Wallace's Giant Bee, an Indonesian species and the world's largest known bee, was presumed extinct after no sightings in the wild since 1981, but it was rediscovered in 2019.

# ENDANGERED INSECTS

*Sadly, many animal species around the world are in danger of extinction. This includes a large number of insects, though their plight is little known compared to larger and more familiar animals.*

Ⓐ Habitat damage and loss harms entire ecosystems, and recovery is a long, slow process.

Wildlife today faces a relentless array of threats, the majority of them caused by human activity. It is no exaggeration to say that a sixth mass extinction is underway, which could prove no less devastating than the last (when the aftermath of a huge asteroid strike destroyed some three-

## Threatened and endangered species

More than a million insects have so far been described by science, and there are likely to be at least five million more that have not yet been discovered. The International Union for Conservation of Nature (IUCN) has so far assessed a tiny proportion of them—nearly 4,300 species. Almost a quarter of those have been found to be at risk of extinction, and there is no reason to suppose that the proportion will change as more species are described and more assessments are completed.

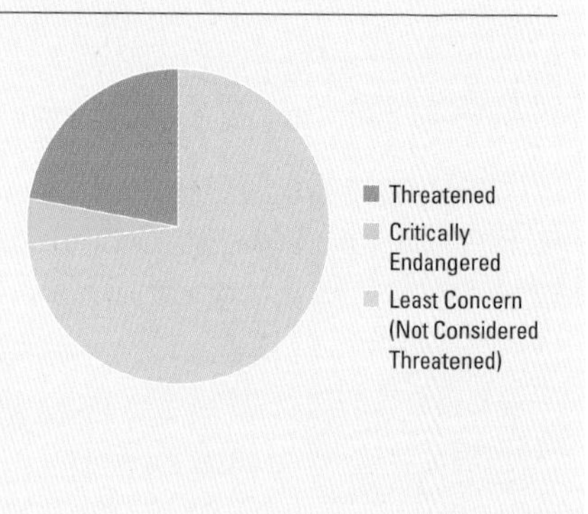

quarters of all plant and animal species on Earth, some 66 million years ago).

The International Union for Conservation of Nature (IUCN) is a global organization that assesses how close each wild species is to extinction, based on all available evidence. The categories it uses are Least Concern (no immediate risk of extinction) through Near Threatened, Vulnerable, Endangered, Critically Endangered, Extinct in the Wild, and Extinct. Species for which insufficient data exists are classed as Data Deficient. At the time of writing, nearly 4,300 species have been fully assessed, of which about 1,150 are placed in one of the threatened categories.

About 200 insect species are classed as Critically Endangered, meaning they are considered to face an extremely high risk of extinction in the wild. Among them are such beautifully named species as the Merry Shadowdamsel Damselfly (*Drepanosticta hilaris*), the Natterer's Longwing Butterfly (*Heliconius nattereri*), and the Canterbury Knobbled Weevil (*Hadramphus tuberculatus*)—but most have no common name and probably never will.

Ⓐ *Heliconious* butterflies from Atlantic forests in Brazil. Documenting the insect fauna of fast-changing habitats is a race against time.

## The true picture?

A further 1,700 insect species are classed as Data Deficient, and a proportion of these are undoubtedly in severe danger too—and then there are unknown numbers of species that have yet to be discovered and described. Many of those will surely die out without ever being known to us.

The rate of species loss around the world today is estimated to be about 1,000 times higher than the natural "background rate." The current mass extinction is occurring across all taxonomic groups, all parts of the world, and in all kinds of habitats, but is especially marked in the most biodiverse parts of the world. Because of insects' ecological importance, their loss has an impact on many other species, from plants to the largest alpha predators.

The knock-on effect of insect extinctions and declines is only too apparent when we look at other animal groups. In North America, for example, there are 2.9 billion fewer birds than there were in 1970 (a decline of 29 percent), and insect-eating species have been disproportionately affected. As we will see on page 215, insect population declines worldwide are also very concerning for humankind.

# INSECT CONSERVATION

**As the 21st century progresses, we are becoming increasingly aware of the wholesale disappearance of wildlife from our planet. Insect biodiversity is plummeting and this impacts all other living things.**

The history of conservation is full of individual success stories, including some that concern insects. One of the best known is that of the Lord Island Stick Insect (*Dryococelus australis*), native to the Lord Howe Island group near New Zealand. This stout phasmid was extinct on Lord Howe Island by 1920, thanks to introduced rats, but in 2001 was found to still survive on a nearby rat-free islet. Two pairs were brought to Melbourne Zoo and a captive breeding program was begun while Lord Howe Island's rats were eradicated. Today, several thousand of the insects have been bred in captivity.

Captive breeding and reintroduction is often successful for restoring populations of particular insects, as long as the factors that caused them to decline in the first place are addressed. In the UK, the Large Blue Butterfly (*Phengaris arion*) has been reestablished from mainland European stock after its disappearance from the UK in 1979, but this project was only possible after biologists had studied the butterfly's ecology and identified the crucial role played by a species of red ant, *Myrmica sabuleti*, in a unique parasitic relationship. Sites for the reintroduction were managed for some years in order to boost ant populations before any of the butterflies were released.

Ⓥ A "bug hotel" like this in your garden will provide nesting and hibernation places for a wide range of invertebrates.

Tropical forest is the most biodiverse land habitat on Earth.

Saving a single, very rare, or range-restricted species from extinction is a worthwhile effort, but in terms of overall ecological impact, just a footnote to the story. It is the general declines, including declines among very abundant species on which many other species depend, that are so damaging to ecosystems as a whole. A 2017 report from Germany, studying data from 63 nature reserves, stated that midsummer insect abundance had fallen by 82 percent between 1989 and 2016. A similar report from the El Yunque National Forest in Puerto Rico reported a loss of 78–98 percent in ground and tree canopy arthropods between 1976 and 2012—these losses are largely attributable to climate change.

Declines on this scale should and do alarm us, not least because a large proportion of the crops we grow depend upon insect pollinators. Other land species are already being affected by the losses—the wild plants that depend on insects to pollinate them, and the animals that feed on insects. Conservation now has to be on a global scale if we are to avert the unfolding crisis.

A reintroduction project brought the Large Blue Butterfly back to England in the 1980s.

## STARTING AT HOME

One of the advantages that insects have is (in most cases) a fast reproductive rate. This means that their numbers can recover dramatically over just a few generations, if conditions improve for them. It is easy to encourage insects to thrive within the little fragments of open space that are our own. Plant your garden with native species, stop using pesticides, and add other natural habitats (perhaps a wildflower meadow, a pond, or a woodpile) to make your own contribution to boosting your local insect fauna. Most countries have at least one charitable conservation body which works to protect insects and their habitats. Joining organizations like these will help support their work and keep you informed about local and national issues affecting insects and other wildlife.

# GLOSSARY

**Abdomen:** the third of the three basic body regions in insects

**Afferent and efferent nerves:** Afferent nerves carry nerve signals from sensory organs to the brain and central nervous system. Efferent nerves work in the other direction, carrying signals from the central nervous system out to other body parts, such as muscles.

**Allele:** One of two or more forms of the same gene, found at the same position on a chromosome. Alleles are responsible for genetic diversity within a species.

**Antennae:** sensory appendages on the head, mainly used to detect odors

**Appendages:** jointed structures attached in pairs to certain body-segments, with variable functions (includes legs and antennae)

**Aquatic:** adapted to live in water

**Cerci:** appendages on the last abdominal segment, often used during copulation

**Chromosomes:** long strands of DNA, each made up of many genes, found in the nucleus of each cell in an animal's body; they come in pairs and all insects of a certain species have the same number of pairs (e.g., the fruit fly *Drosophila melanogaster* has four pairs of chromosomes)

**Chrysalis:** the pupa of a butterfly

**Cocoon:** a protective case of silk around a pupa, produced by some moths and other insects

**Compound eyes:** a pair of large eyes, formed from multiple ommatidia and found on the heads of most insects

**Crop:** the first part of the digestive tract, where swallowed food is stored

**Crop pest:** any plant-eating insect that does significant damage to plants cultivated for food

**Cuticle:** the outer "shell" or exoskeleton of an insect; some internal organs are also cuticle-lined

**Dorsal vessel:** a central vessel that pumps hemolymph, functioning similarly to the heart in vertebrates

**Ecology:** the study of organisms and their environment as a functioning system

**Ecosystem:** the community of organisms within a particular habitat, and their interrelationships

**Elytra:** the modified, thickened forewings ("wing cases") of beetles

**Estivation:** prolonged period of inactivity through hot and dry summer weather

**Eusocial insects:** species that live communally, with most individuals being nonbreeding "workers" acting to maintain the colony and support the breeding female or pair

**Fat-body:** an internal structure where fat is stored; also involved in detoxification and other functions similar to those carried out by the liver in vertebrates

**Ganglia:** a bundle of nerve cells, functioning like a miniature brain

**Genes:** short sections of DNA that each provide the "instructions" for making a particular kind of protein

**Gills:** structures that are used to extract oxygen from water

**Halteres:** club-shaped structures found in Diptera (the true flies), formed from modified hind wings

**Head:** the first of the three basic body regions

**Hemimetabolous:** describes an insect that goes through incomplete metamorphosis, changing from larva to adult with no pupal stage

**Hemocoel:** the internal body cavity of an insect, filled with hemolymph

**Hemolymph:** the fluid that fills an insect's hemocoel, used to transport metabolites

**Hibernation:** passing the winter in an inactive state with greatly lowered metabolism

**Holometabolous:** describes an insect that goes through complete metamorphosis—it passes through four life stages: egg, larva, pupa, and imago/adult

**Imago:** another word for an adult winged insect

**Instar:** growth stages between the molts in an insect larva

**Larva:** an insect that has yet to metamorphize to adulthood; the growth life stage

**Malpighian tubules:** structures connected to the gut that help maintain the correct fluid balance

**Mandibles:** the jaws or biting mouthparts

**Metamorphosis:** transformation from larval form to adult form

**Migration:** traveling (regularly and predictably) from one region to another at a particular time of year, to avoid inclement weather or temperatures

**Mouthparts:** appendages on the head, used for feeding

**Mutation:** During the process of copying chromosomes prior to cell divisions, sometimes errors occur, and the genes that make up the new chromosome copy are not identical to those in the original. This is a genetic mutation, and if it occurs during the formation of a sperm or egg cell that goes on to become an embryo, it will be propagated in all the cells of that embryo.

**Nymph:** an alternative name for the larva of a hemimetabolous insect

**Ocelli:** small simple eyes, found in addition to compound eyes in some insects

**Ommatidia:** the light-sensing structures from which a compound eye is formed

**Parasite:** any animal that lives permanently (in at least one of its life stages) on another animal's body and consumes its body tissues, such as blood or skin

**Parasitoid:** an insect that lays its eggs in or on a living host's body; the larvae will eat the host, eventually killing it

**Predator:** an animal that catches and kills other animals for its food

**Prey:** animals that are killed by predators

**Pupa:** an inactive life stage in between larva and adult in holometabolous insects, during which metamorphosis takes place

**Queen:** a reproductive female in a nest or hive of eusocial insects

**Segments:** the individual body-sections of insects and other arthropods

**Sequestering:** The storing of substances from the diet in the body tissues, in an unchanged state. Some insects sequester toxic compounds from the plants they eat, making their own bodies poisonous to any predators that try to eat them.

**Spiracles:** holes in body-segments, used to take in oxygen and expel carbon dioxide

**Subimago:** a subadult, winged life stage unique to mayflies

**Teneral:** a newly or recently emerged adult insect, usually still soft-bodied and not ready to breed

**Thorax:** the second (middle) of the three basic body regions

**Uniramous:** of a body appendage, having only one branch—either a gill or a leg; some arthropods have biramous (two-branched) appendages, each comprising both a gill and a leg

**Venation:** the branching network of veins within an insect's wing

**Wings:** membranous structures found in most adult insects, and used to provide power and lift for flight

**Worker:** a nonbreeding individual in a eusocial insect colony; its activities may include colony defense, nest maintenance, and collecting food

# INDEX


## A

abdomen 36–7, 38, 94–5, 102
adaptive radiation 18–19
adipocytes 87, 188
aedeagus 108, 110
aeropyles 128, 130, 131
afferent nerves 58
air sacs 94
androconia 43, 116
andromorphs 161
Antarctic Midge 29, 200
antennae 36, 52–3, 54, 99
antlions 139
ants 19, 36, 60, 76, 104, 114, 129, 138, 142, 172, 173, 194, 209
  communication 169
  emergence 153
  honeydew 76, 170, 172, 203
  invasive species 122
  lifespan 144
  nests 62, 207
  pupation 148–9
  social behaviour 168, 169
aorta 98
aphids 114–15, 119, 124, 170, 172, 203, 204
apterygotes 16
aquatic eggs 129
aquatic insects 56, 72–3, 102, 103, 147
aquatic larvae 63, 88, 102, 129, 132, 147, 149, 155
Archaeognatha 13, 18
Arctic Woolly Bear Moth 29, 145, 200
Argentine Ant 122, 207
Asian Lady Beetle 122
Atlas Moth 206
axons 58–9

## B

backswimmers 72
bagworms 133
bark lice 195
bedbugs 121
bees 19, 35, 42, 56–7, 114, 118, 142, 173, 194
  domestication 196
  feeding 158
  flight 70
  intelligence 60
  mating plugs 120
  mouthparts 82
  nests 62
  ocelli 51
  parental care 162–3
  pupation 148–9
  social behaviour 168–9
  stings 111
*see also* honey bees
beetles 18, 34, 80, 115, 142, 170–1, 194, 206, 207
  aquatic 72
  color 46
  elytra 43, 45
  extreme conditions 201
  invasive species 123
  larvae 132, 133
  niches 203
  pests 124
  “playing dead’ 75
biocontrol 204–5
biodiversity 28–9, 30, 62
bioengineering 125
bioindicators 30–1
bioluminescence 207
biomes 198
biosecurity 123
biramous appendages 38
Black Soldier Fly 197
blood-drinkers 83, 88, 159
blowflies 203
body size 11, 24–5, 95, 102, 108, 180, 206
book lice 195
botfly 175
Box Tree Moth 123
braconid wasps 176
brains 58–60
Brimstone Butterfly 144
brood parasites 173
brood sacs 162
Brown Hairstreak Butterfly 128
bumblebees 57, 70, 173
bursa copulatrix 110
burying beetles 171, 172, 203
bush crickets 111, 119, 146
butterflies 19, 35, 43, 46, 54, 116, 118, 138, 142, 172, 194
  conservation 214
  eggs 128
  emergence 155
  extinction 210–11
  honeydew 170
  lifespans 144
  mating plugs 120
  migration 166–7
  molt 136, 137
  proboscis 81, 82
  pupation 148
  salts 88
  startle coloration 74
Butterfly Monitoring Scheme 15

## C

caddisflies 19, 30, 35, 43, 63, 72, 133, 135, 194
camouflage 44–5, 46, 77, 137, 154, 159
cannibalism 118, 131
Carboniferous Period 18, 24–5, 95
carnivores 28, 61, 82–3, 86, 88, 138, 165, 203
cell division 113, 114, 130, 184–5
cell membrane 180–1

cells 178–91
research 190–1
structure 180–1
centrosomes 182, 184
cerci 18, 37, 108, 112, 160
chemoreception 54–5
chitin 34, 94
chorion 128, 129, 130, 131
chromosomes 182, 184–5
chrysalis 149
cicadas 56, 144, 153, 160, 207
circulatory system 39, 92–105
citizen science 14–15
clades 16, 23
"clap and fling' 71
classification 20–3
click beetles 74
climate change 24, 25, 31
climbing 68
cloning 211
cockroaches 17, 18, 25, 68, 80, 162, 194
cocoons 149
collembolans 18
Colorado Potato Beetle 124
colors 46–7, 74
commensalism 170–1
Common Froghopper 206
communication 60, 169
complete metamorphosis 88, 138, 143
compound eyes 36, 50–1
conservation 192–215
cooperative behaviour 60
coremata 45
cornea 50–1
corpora pedunculata 169
costa 42
Cotton Aphid 124
courtship displays 43, 53, 112, 116–17, 160–1
coxa 38
Cream-spotted Tigerwing 149
crickets 56, 68, 80, 132, 142, 160, 194
crop 84, 85, 169
crustaceans 20, 38
cuckoo bumblebees 173
cuticle 34, 40, 46, 54, 85, 90, 101, 102, 136, 147, 148, 152, 186
cytoplasm 180–3

## D

damselflies 16, 22, 31, 143, 160–1, 194
extinction 211
flight 70
mating 112, 160
mouthparts 81, 88
prolarvae 143, 146
setae 40
sperm competition 109
teneral 154
DDT 30–1, 208–9
demoiselle damselflies 160
dendrites 59
digestion 78–91
digestive tract 84–5
diplurans 18
disease 187
diving 56, 72–3
diving beetles 40, 73, 103, 132, 134–5
DNA 20, 182, 211
domesticated insects 196–7
dragonflies 17, 18, 22, 24–7, 46, 72, 103, 136, 142, 159, 194, 206
climate change 31
eyes 50, 51
flight 16, 70
hemolymph 102
intelligence 61
lifespan 144–5
mating 37, 112
migration 166, 167
mouthparts 34, 81, 83, 88
niches 203
obelisking 165
setae 40
sperm competition 109
teneral 154
territory 116
drinking 90–1
dung beetles 117, 118

## E

earwigs 162, 194, 211
ecology 14
ecosystems 202–3
efferent nerves 58
eggs 108, 110–11, 112–15, 118–19, 126–39, 161, 177, 185, 191
ovipositors 54–5, 110–11, 119
parental care of 163
size 147
electrical fields 57
elytra 43, 45, 73, 103
emergence 152–5
endangered species 212–13
endocrine system 104–5
endoplasmic reticulum 182
enzymes 183, 207
epicuticle 34, 46
epidermis 34, 35, 40, 189
estivation 144, 165
eurypterids 11
eusociality 168–9, 206
everted body parts 45
evolution 8–31, 143
excretion 85, 91, 150
exocrine glands 189
exocuticle 34
exoskeleton 10–11, 34–5, 95, 102, 136
extinction 210–12, 213
Eyed Hawkmoth 74
eyes 36, 50–1

# INDEX


## F

fairy longhorn moths 53
Fall Armyworm Moth 191
feeding 78–91
  behaviour 158–9
  larvae 134–5, 138–9
female choice 112, 116, 117, 120–1, 160
femur 38–9, 68
flagellomeres 52
flagellum 52
flatworms 84
fleas 19, 27, 83, 152, 174
flight 16–17, 42–3, 70–1
fluid balance 90–1
fossils 10, 12–13, 20
froghoppers 206
fruit flies 102, 120, 151, 197,

## G

ganglia 58–9
ganglion cells 185
gardens 198, 204, 215
gas exchange 96–7
gene sequences 20
genetics 151, 197, 211
germ band 130
Giant Water Bug 163
Giant Weta 180, 206
Giraffe Weevil 44
Glasshouse Whitefly 204
Glassy-winged Sharpshooter 204
glial cells 185, 188
global distribution 28–9
Globe Skimmer Dragonfly 206
Golgi apparatus 182
grasshoppers 17, 42, 56, 68, 80, 113, 132, 142, 160, 194
Green Darner Dragonfly 167
griffinflies 24–5
Gypsy Moth 134, 187

## H

habitat 198–200
  loss 208
hair *see* setae
halteres 43
hawker dragonflies 145
hawkmoths 148, 158, 206
head 34, 36–7, 38, 58
Head-stander Beetle 201
Headlight Elator Beetle 207
hearing 56
hemelytra 43
hemimetabolous insects 142, 143, 146, 147, 148, 152, 194
hemocoel 98–9
hemocyanins 100
hemocytes 101, 186
hemolymph 85, 87, 90, 91, 98–102, 104, 136, 147, 152, 183, 186–7
herbivores 28, 61, 82, 86, 89, 123–5, 132, 138, 165, 198, 203
Hercules Beetle 44, 206
hexapods 18
hibernation 77, 144, 164–5, 204
holometabolous insects 142–3, 147, 148, 149, 152, 194
honey bees 56–7, 60, 114, 135, 149, 154–5, 161
  cells 180
  domestication 196
  immune response 187
  social behaviour 154, 168–9
honeydew 76, 170, 172, 203
hormones 104–5, 136, 146, 181
horseflies 83, 159
houseflies 69, 90
hoverflies 103, 204
hox genes 151
Hymenoptera 19
hypopharynx 80

## I

ichneumon wasps 176
imaginal disks 150
immune response 183, 186–7
incomplete metamorphosis 88, 132, 137, 142, 146–7
instars 136–7, 146–7
intelligence 60–1
intima 85
invasive species 122–3, 167, 204, 209
iridescence 46

## J

Japanese Red Wood Ant 207
Johnston's organ 52

## K

katydids 194
keystone species 203

## L

labium 80, 81, 82, 88, 148
labrum 80
lacewings 72, 118–19, 133, 194, 204
ladybugs 122, 132, 165, 204
Large Blue Butterfly 172, 214
Large White Butterfly 176
larvae 88, 118–19, 126–39, 142, 146–8
  aquatic 63, 88, 102, 129, 132, 147, 149, 155
  hatching 131
  hemocytes 101
  hemolymph 102
  hormones 104, 105
  mouthparts 81, 88
  paedogenesis 115
  parasitoids 176–7

parental care of 162–3
plant toxins 89
leaf insects 45
leaf-mining moths 134
leaf rollers 63
leafhoppers 204
leaping 68, 206
legs 36–7, 38–9, 68
muscles 66, 67
structure 40–1
swimming 72
lekking 117
*Leptinotarsa decemlineata* 124
lice 174
life cycles 88–9, 142–5
lifespans 71, 144–5
lily beetles 133
Lily Leaf Beetle 123
locusts 124, 138–9, 211
longhorn beetles 52
Lord Howe Island Stick Insect 117, 214
louse flies 27
lysosome 183


## M


magnetic fields 56–7
malaria 26, 175
Malpighian tubules 84–5, 91
mandibles 80–1, 82–3, 86, 134, 152, 159
mantises 83, 120, 142, 194
mass extinctions 11, 212, 213
mating 112–13, 116–17, 120, 160–1
mating plugs 120
maxillae 80–1, 82
mayflies 16, 17, 18, 25, 30, 42, 70, 72, 103, 142
emergence 153
larvae 129, 132
molt 155
meiosis 185
mesothorax 36
metamorphosis 88, 132, 137, 138, 140–55
metathorax 36
micropyle 113, 128
microscopy 190
midges 115
Migrant Hawker Dragonfly 31
migration 165, 166–7, 206
mimicry 159
mitochondria 182, 183, 189
mitosis 184–5
mole crickets 40
molt 34, 90, 100–1, 104, 105, 136–7, 142, 143, 146–7, 148, 155
Monarch Butterfly 166, 167
mosquitoes 26, 83, 88, 149, 175, 188
moths 19, 35, 43, 54, 142, 148, 165, 187, 194, 206
antennae 52–3
aquatic larvae 72
caterpillars 27, 35, 63, 105, 133, 134, 200
cellular research 191
*Creatonotos gangis* 45
domestication 196–7
emergence 155
extreme conditions 200
feeding 158
flightless 76
*Gynaephora groenlandica* 29, 145, 200
hearing 56
hemolymph 102
invasive species 123
lifespan 145
pests 125
proboscis 81, 82, 158
pupae 150
startle coloration 74
mouthparts 36, 34, 54, 80–3, 84, 86, 88
muscles 66–7, 183
cells 188–9
flight 70
mutualism 170–1
myocytes 66, 188–9


## N


neck (cervix) 37
nectar 57, 60, 82, 83, 88, 135, 139, 154, 158, 163, 168–9, 198
nerves 58–60, 181, 185, 188
nervous system 39, 48–63
nests and shelters 62–3, 132–3, 148–9, 162–3, 164, 169, 171, 204
neurotransmitters 181
niches 203
nucleus 182
nymphs 132, 138


## O


obelisking 165
ocelli 51
olfactory communication 169
ommatidium 50–1
omnivores 82–3, 203
orchid bees 117
orchid mantises 45, 159
organelles 181, 182–3, 184
osmotic pressure 90
ostia 98–9
ova *see* eggs
ovaries 110
ovipositors 54–5, 110–11, 119


## P


paedogenesis 115
Painted Lady Butterfly 166–7
pair bonds 117
palaeodictyopterids 17, 25
palps 80, 81
parasitic insects 27, 83, 119, 172–5, 195, 203

# INDEX


parasitoid wasps 111, 172, 176–7, 186, 204
parasitoids 149, 176–7
Parent Bug 162
parental care 116, 162–3
parthenogenesis 114–15
pedicel 52
Permian Period 11, 17, 26
pest control 105, 124–5, 134, 139, 155, 174, 187, 204–5, 208–9
phagocytosis 101, 186
phasmids 45, 128–9, 214
pheromones 45, 53, 155, 172, 189
phoresy 172
photoreceptors 50–1
picture-winged flies 160
pigmentation 46
plant toxins 89, 134
plastron 73, 129
"playing dead" 75
polar regions 29
pollen baskets 35
pollination 27
pollution 30–1
potter wasps 176–7
praying mantises 40
proboscis 81, 82, 150, 158
procuticle 34
prolarvae 143, 146
prolegs 102
pronucleus 113
prothorax 36
proturans 18
proventriculus 84, 86
pruinescence 46
pseudocone 50
pterygotes 16
pupae 148–52
pupal mating 155
pupation 148–9, 150

## R

reproduction 106–25
  female system 110–11
  male system 108–9
  reproductive organs 37, 108–11
  *see also* eggs; sperm
respiration 39, 92–105
rhabdom 50
rhinoceros beetles 44
*Rhyniognatha* 12, 13
ribosomes 182
Ringlet Butterfly 118
Riverfly Monitoring Initiative 30
robber flies 159
Rocky Mountain Parnassian Butterfly 120
rove beetles 45
running 40, 68, 206

## S

saliva 84, 86
sawflies 19, 119, 132, 194
scale insects 76
scales 35, 43, 46, 116
scape 52
scarab beetles 40
scientific names 23
sclerotization 34, 43, 44
scorpionflies 18–19, 158
seasonal behavior 164–5
segments 38–9, 40
senses 48–63
sensilla 54, 56
septa 99
serosa 130
serotonin 138
sessile insects 76–7
setae (hair) 35, 40, 43, 69, 72, 73
shield bugs 118, 162
Silk Moth 104, 196–7
Silver-Washed Fritillary Butterfly 116
silverfish 13, 112–13, 142
smell 54–5
social structure 60
song 160, 207
specialization 26–7, 28, 82–3
species diversity 192–215
species numbers 28–9
sperm 88, 108–9, 110, 112–14, 120–1, 183, 185, 188, 191
sperm competition 109
spermathecae 110
spermatophores 113
spiracles 90, 94, 95, 96, 186
Spotted Stem Borer Moth 125
startle coloration 74
stick insects 114, 194, 206, 214
stings 111
stone flies 30, 72, 100, 142, 194
structural coloration 46
subimago 155
Swallowtail Butterfly 137
Swift Lousefly 201
swimming 40, 72–3
synapses 59

## T

tarsus 39, 40, 54, 68
taste 54–5
taxonomy 20–3
temperature
  extreme conditions 200–1
  range 164
  regulation 102, 165
teneral insects 154
termites 62, 104, 144, 168, 169, 171, 194
territory 116–17
testes 108

thorax 16–17, 36–7, 38, 58, 70, 74, 94, 102
thrips 71
tibia 39, 40
tiger beetles 68, 206
touch 56
trachea 24, 94, 96, 102
trap-lining 158
traumatic insemination 121
trigonotarbids 11
trilobites 10–11
trochanter 38
true bugs 25, 43, 72, 76, 80–1, 82, 142, 170, 194, 198
true flies 19, 43, 54, 72, 83, 102, 132, 142, 150, 194
tsetse flies 119
tymbals 160
tympanal organ 56

## U

ultraviolet light 46, 51
uniramous appendages 36, 38
uric acid 85, 91, 100

## V

Vagrant Emperor Dragonfly 144, 166
Vapourer Moth 76–7
ventriculus 84
vision *see* eyes
volunteers 14–15, 30

## W

waggle dance 60, 169
walking 68
  upside down 69
wasps 19, 27, 36, 42, 114, 118, 135, 142, 194
  feeding 159
  nests 62
  ocelli 51
  parental care 162–3
  pupation 148–9
  social behaviour 135, 168
  stings 111
  *see also* parasitoid wasps
water boatmen 72
water pollution 30
water striders 73
web spinners 195
webs 63
weevils 18
whirligig beetles 72, 206
White Witch Moth 206
wings 17, 36, 42–3, 70–1, 99, 137, 147, 152–3
  courtship 160
  scales 35, 43, 46, 116
  temperature regulation 165
wood-boring beetles 170–1

## X

Xerces Blue butterfly 210–11

## Y

Yellow Crazy Ant 209

## Z

Zygentoma 13, 18

# PHOTO CREDITS

**Alamy**
13 lower © The Natural History Museum / Alamy Stock Photo
91 © Cultura Creative (RF) / Alamy Stock Photo
211 lower © Nicolas Vereecken / Alamy Stock Photo

**Creative Commons**
19 upper © Gregor Nidar
87 © Beatriz Moisset
43 both © Sharp Photography
154 lower © Stemonitis
177 lower © Hectonichus
211 upper © Brianwray26

**Nature Picture Library**
95 © Tim Edwards
151 © Radu Razvan
154 upper © Paul Cowan
163 left © John Cancalosi
165 left © Philippe Clement
185 © Visuals Unlimited
203 © Jan Hamrsky

**Science Photo Library**
25 © RICHARD BIZLEY
29 © LOUISE MURRAY
69 upper © Stephen Dalton

**Shutterstock**
9 © Vladimir Prokop
10 © Merlin74
12 © Hhelene
13 upper © Marco Uliana
14 © Ann in the uk
15 upper © Jacky D
15 lower © Matt Jeppson
17 upper © gan chaonan
17 lower © Pavel Krasensky
19 lower © FerencSpeder84
23 © macgorka
23 © Gerald A. DeBoer
26 © John Gregory
27 upper © Pavel Krasensky
27 lower © Vlasto Opatovsky
29 © 2seven9
30 © Alexmalexra
31 upper © Mirko Graul
31 lower © MD_Photography
33 © alslutsky
34 left © hwongcc
34 © Eric Isselee
35 © Tomatito
36 © Daniel Prudek
37 upper left © Tomatito
37 upper right © Costea Andrea M
37 lower left © Eileen Kumpf
37 lower right © Henrik Larsson
38 both © Peter Halasz
39 left © Sandra Standbridge
39 right © Mospan Ihor
40 © Johan Larson
41 upper left © Bildagentur Zoonar GmbH
41 upper right © Sriyana
41 middle left © OlegD
41 middle right © Cornel Constantin
41 lower right © MityaC13
42 top two © Paul Stout
42 middle © Violetta Honkisz
42 lower-mid © symbiot
42 bottom © Aksenova Natalya
44 upper © Leisha Kemp-Walker
44 lower © Alen thien
45 © Lightboxx
47 top left © szefei
47 top right © Elena Elisseeva
47 upper-mid left © Klaas Vledder
47 upper-mid right © ninii
47 lower-mid left © Mark Brandon
47 lower-mid right © Henri Koskinen
47 bottom left © yxowert
47 bottom right © kristine.tanne
49 © Skyler Ewing
50 © finchfocus
51 © Tomasz Kowalski
52 upper © IanRedding
52 middle © Martin Pelanek
52 lower © Ttstudio
53 left © PhotoAlto/Christophe Lemieux
53 right © Rudmer Zwerver
55 upper © gardenlife
55 lower left © Anteromite
55 lower right © Lutsenko_ Oleksandr
56 © Radu Bercan
57 © Jari Sokka
59 © Irina Kozorog
60 © frank60
61 upper © Dancestrokes
61 middle © Ishwar Thakkar
61 lower © corlaffra
62 upper © monika3steps
62 lower © Erika Bisbocci
63 left © travelpeter
63 right © scubaluna
65 © Evgeniy Melnikov
67 © Kevin Wells Photography
68 upper © Eric Isselee
68 lower © Alexander Sviridov
69 lower © Anteromite
71 upper © Poppap pongsakorn
71 lower © Tomasz Klejdysz
72 left © lehic
72 right © Henrik Larsson
73 upper © Jan Miko
73 lower © Martin Pelanek
74 © Pyma
75 © SIMON SHIM
76 © Timothy Weinell
77 upper © thatmacroguy
77 lower © Protasov AN
79 © Tin21
81 upper © Cornel Constantin
81 lower © narong sutinkham
82 © Ernest Cooper
83 upper © Photo Insecta
83 middle © Henrik Larsson
83 lower © 108MotionBG
85 upper © Afanasiev Andrii
85 lower © Kazakov Maksim
88 left © Jolanda Aalbers
88 right © Vitalii Hulai
89 upper left © Savo Ilic
89 upper right © Randy Bjorklund
89 lower © AjayTvm
93 © Alexey Lobanov
94 © Protasov AN
97 © AlessandroZocc
99 upper © Henrik Larsson
99 lower © Ezume Images
100 © Darkdiamond67
102 © Christian Musat
103 upper © Marco Maggesi
103 lower © Rainer Fuhrmann
105 upper © Creatikon Studio
105 lower © Naronta
107 © lucky vectorstudio
109 upper three © Kuttelvaserova Stuchelova
109 bottom © Kluciar Ivan
111 upper left © Helen Cradduck
111 upper right © Brett Hondow
111 lower left © Perry Correll
111 lower right © Eileen Kumpf
112 © Geza Farkas
113 left © Premaphotos
113 right © Isti yano
114 © Tomasz Klejdysz
115 © thatmacroguy
116 © Alta Oosthuizen
117 left © Sanjoy Krishna Das
117 © Phil Savoie
118 © Predrague
119 left © aaltair
119 right © Holger Kirk
120 © DM Larson
121 upper © Eataru Photographer
121 lower © Judith Lee
122 © IanRedding
123 © Michael W NZ
123 upper left © JOTAQUI
123 upper right © Javier Chiavone
124 © Rupinder singh 0071
125 upper © Lost Mountain Studio
125 middle © Tomasz Klejdysz
125 lower © Dario Sabljak
127 © Nicole Patience
128 © Eric Isselee
129 upper four © Drägüs
129 lower © Stephan Morris
130 © Cathy Keifer
131 © Sugi8816
132 © SIMON SHIM
133 upper © Lawrence Jefferson
133 lower © Helissa Grundemann
134 lower © FAGRI
135 upper © Michael Benard
135 middle © Richard Bartz
135 lower © Witsawat.S
137 © Joaquin Corbalan P
138 © Margus Vilbas
139 upper © Charlotte Bleijenberg
139 middle © Tim Lamprey
139 lower © leandros soilemezidis
141 © muhamad mizan bin ngateni
143 © eleonimages
144 © wonderisland
145 © Gerry Bishop
145 © JoannaGiovanna
146 upper © thatmacroguy
146 lower three © Michal Hykel
147 lower three © Michal Hykel
147 © watthana promsena
148 © Linda Bestwick
149 © Igor Chus
150 © de2marco
152 all © halimqd
153 © Danuphon Leesuksert
157 © Jiri Hrebicek
158 left © Sarah2
158 right © Sriyana
159 left © Jiri Hrebicek
159 right © Codega
160 © Martina_L
161 upper © Jolanda Aalbers
161 lower © JEONG MANHEUNG
162 © Vinicius R. Souza
163 right © Kazakov Maksim
164 upper © Illiyin
164 lower © Vitalii Hulai
165 right © Rainer Fuhrmann
166 © Danita Delmont
167 upper © tony mills
167 lower © Jeff Holcombe
168 upper © Simun Ascic
168 lower © Elen Marlen
169 © MicrotechOz
170 © Gerhard Prinsloo
171 upper © Stephen Barnes
171 lower © IanRedding
172 © kallen1979
173 upper © Milan Rybar
173 lower © Predrague
174 left © Tacio Philip Sansonovski
174 right © lucasfaramiglio
175 left © Protasov AN
175 right © photowind
176 © Jumos
177 upper © Nik Bruining
179 © Viktor Gladkov
181 upper © Oleg_Serkiz
181 lower © Dominic Gentilcore
182 © Guillermo Guerao Serra
183 © Jose Luis Calvo
187 © Nick Upton
189 © Nenad Nedomacki
190 © McElroy Art
191 © metita ungsumeth
193 © Oleksandrum
194 © Henri Koskinen
195 upper left © Dmitry Monastyrskiy
195 upper right © Lutsenko_ Oleksandr
195 lower © Paco Moreno
196 © santypan
197 upper left © ABS Natural History
197 upper right © Ilizia
197 lower © Studiotouch
199 upper left © NatchaS
199 upper right © Damsea
199 mid-right © photos_adil
199 lower left © CHANNEL M2
199 lower right © Kunlaphat Raksakul
199 mid-left © rmk2112
200 © Pavel Krasensky
201 upper © fritz16
201 lower © Piotr Velixar
205 upper right © Anthony Fergus
205 lower © dohtor
206 © Cocos.Bounty
207 lower © Georges_Creations
207 © Salparadis
208–209 © Ed Goodacre
208 upper © Young Swee Ming
212 © Tarcisio Schnaider
214 © dvlcom
215 upper © Jhurst84
215 lower © Cesar J. Pollo

**Illustrations by Robert Brandt**